Michael Faraday: A Very Short Introduction

VERY SHORT INTRODUCTIONS are for anyone wanting a stimulating and accessible way in to a new subject. They are written by experts, and have been published in more than 25 languages worldwide.

The series began in 1995, and now represents a wide variety of topics in history, philosophy, religion, science, and the humanities. The VSI Library now contains over 200 volumes—a Very Short Introduction to everything from ancient Egypt and Indian philosophy to conceptual art and cosmology—and will continue to grow to a library of around 300 titles.

Very Short Introductions available now:

For more information visit our web site
www.oup.co.uk/general/vsi/

Frank A. J. L. James

MICHAEL FARADAY

A Very Short Introduction

OXFORD
UNIVERSITY PRESS

OXFORD
UNIVERSITY PRESS

Great Clarendon Street, Oxford ox2 6DP
Oxford University Press is a department of the University of Oxford.
It furthers the University's objective of excellence in research, scholarship,
and education by publishing worldwide in

Oxford New York

Auckland Cape Town Dar es Salaam Hong Kong Karachi
Kuala Lumpur Madrid Melbourne Mexico City Nairobi
New Delhi Shanghai Taipei Toronto

With offices in

Argentina Austria Brazil Chile Czech Republic France Greece
Guatemala Hungary Italy Japan Poland Portugal Singapore
South Korea Switzerland Thailand Turkey Ukraine Vietnam

Oxford is a registered trade mark of Oxford University Press
in the UK and in certain other countries

Published in the United States
by Oxford University Press Inc., New York

British Library Cataloguing in Publication Data

Data available

Library of Congress Cataloging in Publication Data

Data available

Typeset by SPI Publisher Services, Pondicherry, India

Printed and bound by
CPI Group (UK) Ltd, Croydon, CR0 4YY

ISBN 978-0-19-957431-5

In Memoriam
David Gooding
(1947–2009)

Requiescat in pace

David died, after a long and painful illness, just two days before this book was to be sent to the publishers. Instead of redrafting the Preface, I am including this dedication page as a tribute to one of the finest scholars I have ever known.

FAJLJ
Isleworth, 13 December 2009

Contents

Preface

Having worked on Faraday for more than a quarter of a century, I have accumulated enormous debts to scholars, archivists, librarians, and curators as well as to my friends and colleagues at the Royal Institution. All these are recorded in detail at the beginning of each volume of my edition of Faraday's *Correspondence*, but I cannot publish this *Very Short Introduction* without acknowledging those to whom I owe most. In terms of Faraday scholarship, David Gooding, Geoffrey Cantor, and Ryan Tweney must take pride of place for endless discussions on Faraday and for their friendship and support over many years. Some of the results of all this can be seen from the frequency with which their names occur in the Further Reading section. Cantor also very kindly read the entire typescript and made many useful comments and suggestions. Away from Faraday studies, I must pay tribute to Marie and Rupert Hall, who sadly both died in February; the experience they provided at Imperial College taught me (and many others) to think historically. One of those fellow students was J. V. Field, who was, as ever, happy to listen and provide a sounding board as writing proceeded. Nevertheless, the usual authorial health warnings apply.

This *Introduction* is firmly rooted in the 250 or so archives around the world that hold Faraday's manuscripts. Particular thanks for permission to work on their collections must go to the main holders, namely the Royal Institution (and for permission to publish Figures 2, 3, 6, 7, 8, 9, 12, 13, 20, 21, and 22), the Institution of Engineering and Technology (and for permission to publish Figure 23), the Guildhall Library, and the Royal Society. For this book, the Cumbria Record Office at Kendal (formerly the Westmorland Record Office), the special collections of Lancaster University (for the records of Clapham Wood Hall), the National Archives, the Geological Society, Dundee University Archives (and for permission to publish Figure 4), and the Camden Local Studies and Archives Centre (for the records of Highgate Cemetery) provided valuable pieces of information on Faraday's history.

Finally, almost invariably for prefaces, but no less heartfelt for that, I must thank my wife, mother-in-law, and children for all their enormous support and interest over the years. Such was the influence of Faraday in our home that when the children were very young, they were under the impression that Faraday was the last day of the working week – fame indeed.

FAJLJ
Royal Institution, London

List of Illustrations

Introduction

This book is a history of Michael Faraday from the 18th century through to the early 21st rather than a conventional biography. An individual's life tempts biographers to concentrate on an apparently well-defined entity, with a straightforward chronological narrative spanning the cradle to the grave, dealing simply with their subject but paying little regard to events elsewhere in their world. A history of a life seeks to place the subject in the context of their times, their politics, their culture, and, in this case, their science and religion. Furthermore, few biographies deal at any length with the reputation of their subject after death. In Faraday's case, it is essential to understand his fame and celebrity, both before and after his death, because the image his 19th-century biographers bequeathed is remarkably at variance with the Faraday uncovered by recent historical study.

Modern scientists and engineers, following the early biographies, still see Faraday's significance in his important experimental discoveries such as benzene (1825), electro-magnetic induction (1831), electro-chemistry (early 1830s), the magneto-optical effect and diamagnetism (both 1845), and thereafter, in consequence, the establishment of that cornerstone of modern physics, the field theory of electro-magnetism. More recently, his colloid work (1856) has, quite anachronistically, been seen as the beginnings of nanoscience. For historians, on the contrary, Faraday's history provides excellent material for investigating how science related to other areas of 19th-century society and culture. This also provides entry points to understand how Faraday was seen by himself and others.

Since this book is an introduction, it examines broad relationships between different areas of Faraday's history, rather than delving into specific details unnecessarily. Thus only two chapters discuss his scientific research, whilst equal space is devoted to his reputation. The opening chapters deal with his beginnings, institutional context, and practical work. Themes that permeate the chapters include his religion, his approach to studying the natural world, his income, the role of Britain in the world, the influence of living and working in London, and his position in society.

In many ways, Faraday can be seen as a pillar of the establishment, undertaking an enormous amount of work for the state and its agencies and being rewarded with a Civil List pension and a grace and favour house. But, by belonging to the Sandemanians, a small sect of literalist Christians, founded in direct opposition to the state church, he was significantly outside the establishment. This meant, for example, that he played only a minor role in the British Association for the Advancement of Science run by Anglican gentlemen of science, although he was on good terms with many of them, including those in holy orders. His faith was central to the practice of his science in terms of the knowledge he produced and how he communicated and applied it.

Faraday's science addressed these three themes – research, communication, application – a categorization he would have recognized. All were linked in various ways during his lifetime, although the proportions of time he devoted to each changed over the years. To pursue research, he gave up a significant income derived from providing commercial and legal scientific advice. Furthermore, to maximize his time for research, he deliberately distanced himself from many of the scientific community's social norms; in almost invariably declining invitations to dinner, in not wishing to be knighted, and refusing the Royal Society Presidency (twice), he went directly against prevailing scientific mores.

Born shortly before the final war between Britain and France commenced, Faraday was 24 when peace finally returned to Europe and Britain emerged as the world's dominant power, with a vast, if diverse, empire that expanded throughout his life. While much of this book locates Faraday in his immediate London contexts, we need to remember his broader European and imperial connections which contributed significantly to his fame.

Because of its length, the war severed the strong relationships that had existed between savants in Britain and on the Continent during the 18th century, and Faraday worked hard developing his own contacts outside Britain. He successfully formed close friendships with scientific figures such as Christian Schoenbein (of Basle), Julius Plücker (Bonn), Justus Liebig (Giessen and Munich), J. N. P. Hachette (Paris), Joseph Plateau (Ghent), J. B. A. Dumas (Paris), Macedonio Melloni (Naples), Auguste De La Rive (Geneva), Carlo Matteucci (Paris and Pisa), and others, such as Alexander Humboldt (Berlin), André-Marie Ampère (Paris), D. F. J. Arago (Paris), Gerard Moll (Utrecht), Moritz Jacobi (St Petersburg), and J. J. Berzelius (Stockholm), whom he knew reasonably well.

In order to exploit and sustain Britain's empire, scientific knowledge was urgently needed. This was one of the reasons why the Royal Institution was founded in 1799 by the President of the Royal Society, Joseph Banks. Banks wanted an institution that would promote the practical value and use of science in agriculture, industry, and, especially, the empire. This was not the role for a traditional learned society, such as the Royal Society.

Applying science imperially was not a new theme in his thinking: he had proposed the ill-fated voyage of HMS *Bounty* in the late 1780s to take breadfruit plants from the East to the West Indies.

The Royal Institution achieved Banks's aims by providing lectures for a general audience and scientific advice to those who required it. Faraday, working there virtually all his adult life, happily continued this agenda. He became the most popular lecturer in London during the middle third of the century and advised on a whole range of topics from chemical analysis of water 'for the good of Australia' to helping design colonial lighthouses. All this was highly political, but he seldom commented on politics, though he admired 'that true old English Gentleman', the radical politician Francis Burdett sent to the Tower in 1810. In any case, lacking property, Faraday did not have a vote.

Faraday's estate, as he noted, was time. Born poor, he received a standard working-class education, did not marry money, and therefore needed to make his living through using his time wisely. Much of his income in his early years at the Royal Institution was produced by doing legal and commercial chemical analyses (including analysing gunpowders for the Honourable East India Company) which he estimated earned him £600 annually, although his successor, John Tyndall (who had access to his account books, now lost), noted earnings of more than £1,000 in 1832. As his income from other sources increased during the 1830s, he gradually ceased undertaking this kind of work in order to concentrate on scientific research. After 1836, he never earned less than £800 annually, and his will was proved at just under £6,000.

Never formally trained in science (few had been), Faraday developed his own methodology to explore experimentally and interpret natural phenomena non-mathematically, although highly spatial and visual in practice. In many ways, Faraday can be

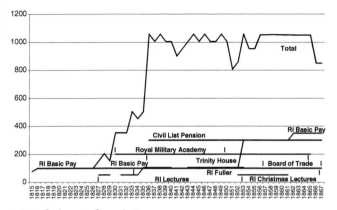

1. Faraday's annual earnings

seen in the tradition of English empiricism going back to
John Locke, but he was quite capable of ignoring, at least
temporarily, experimental results if they contradicted his view
of the universe.

Faraday's way of pursuing science displayed his obsessive character
and his persistent attention to detail, returning to problems,
occasionally over decades, until the experiments produced the
results he wanted. He enabled himself to do this by constructing
a powerful information-retrieval system. In 1832, he started
numbering the paragraphs of his laboratory notebooks and
28 years later ended with paragraph 16,041. Every so often he
would go back and index these notes and other documents that he
had so numbered (such as the 3,430 paragraphs in the 32 papers,
published mostly in the Royal Society's *Philosophical
Transactions*, of his series 'Experimental Researches in Electricity'
and the 1,282 paragraphs of *Chemical Manipulation* (1827),
his only book conceived as such). It was his methodology,
obsessiveness, and persistence that were crucial in allowing him
to develop and articulate his understanding of the world. That he
could do this depended on the contingencies of his history.

Chapter 1
Beginnings

The Faraday family

Presumably conceived around Christmastime the previous year, Faraday first saw the world to which he would contribute so significantly on Thursday, 22 September 1791, a fine autumn day. He was born in Newington Butts, Southwark, just over a mile south of Blackfriars Bridge, one of the three fixed crossings of the Thames. Newington Butts was then at the edge of London; beyond lay the fields of Surrey, and indeed as late as the 1820s farm animals were a common sight there. The London into which Faraday was born was then the largest, richest, most powerful city in the world; George III was King and William Pitt Prime Minister. But London could still be crossed on foot from west to east in a little over an hour and far more quickly north to south. Although Faraday came to enjoy an international reputation, his life and career were firmly rooted in London. As his parents were not Londoners, it is essential to understand the contingencies that led to his birth there.

Although the Quarter day of Michaelmas was only a week away, Faraday was probably named after his maternal grandfather, Michael Hastwell, and this name was recorded in the family Bible. Little is known about Hastwell, other than that during the middle part of the 18th century he was tenant of Black Scar Farm at Kaber to the north of Kirkby Stephen, a market town at the head of the Eden Valley in what was then Westmorland in north-west England. He probably married in the early 1750s, the name of his wife was possibly Betty, but we do know that they had ten children. In 1777, one of their daughters, Mary, married, in the parish church of Kirkby Stephen, Richard Faraday who had moved from Clapham in the West Riding of Yorkshire, some 30 miles to the south. Richard Faraday's parents, Robert Faraday and Elizabeth Dean, had married in 1756 and lived in Clapham Wood Hall which they co-owned with Elizabeth's two sisters. Robert Faraday farmed this smallholding of about 46 acres which also had a small water-driven weaving mill. He and Elizabeth had seven sons and three daughters, the last born in 1776.

Clapham Wood Hall was not capable of supporting ten children as they grew to maturity, especially as none of the daughters married until they were in their thirties, during the first decade of the 19th century. In any case, sons were expected to make their own way in the world. Quite why Richard chose to move to Kirkby Stephen is not known; nevertheless, he prospered there and by the time of his death in 1815 he owned two houses and two mills in the town. His success there doubtless accounts for one of his younger brothers, James Faraday, moving at some point to be a blacksmith at Outhgill five miles south of Kirkby Stephen.

There he courted Margaret Hastwell, a sister of his brother's wife. Conveniently, she was in service at Deep Gill Farm on the other side of the Eden Valley from Outhgill and they married in Kirkby Stephen parish church on 11 June 1786, four days after James's father's death. At some point shortly afterwards, they moved to London, probably prompted by the demography of Westmorland. Always the least densely populated county in England (54 people per square mile in 1700), Westmorland's population increased by nearly 3,000 during the 18th century to just under 43,000 in 1801, in addition to which just over 5,500 people migrated from the county to elsewhere in Britain between 1781 and 1801.

James and Margaret Faraday settled in Newington Butts, where they quickly had two children, Elizabeth (born 26 May 1787) and Robert (8 October 1788), both named after their father's parents. On 9 November 1788, Elizabeth and Robert were baptized in the Anglican parish church of St Mary, Newington. Michael was born nearly three years later. In the mid-1790s, the Faradays moved to rooms over a coach house in Jacob's Mews near Manchester Square on the western edge of London but close to Welbeck Street where James Faraday worked for the ironmonger James Boyd; they remained there until 1809.

Faraday's formative years from the age of 5 to 14 were thus spent in Jacob's Mews. He attended a 'common day-school' where he later recalled that 'my education was of the most ordinary description, consisting of little more that the rudiments of reading, writing, and arithmetic'. A glimpse of his life at that time is caught in a letter written nearly 60 years later to a nephew, describing the building of the canal at Paddington, opened in 1801, and about a dozen streets away from Jacob's Mews:

> When they first formed the canal at Paddington then a set of desolate or field like country it interested me very much to see how they went about their work. I was only a boy but the workmen allowed the boys to run about amongst them and it was to us so strange to us to see a stagnant shore found where we had been used to play. Then I remember also that after a little while the first ever barge appeared there. Not an iron barge only but one carrying goods & coals & I learnt at that time some of my philosophy and set various things beside iron saucepans & pot[s] afloat – for jugs bottles & many other things that I thought at first ought to sink floated & so many first steps to knowledge were gained[.]

Most children have an inquisitive nature which they tend to lose as they grow; Faraday was unusual in keeping his.

James Faraday's mother died in early 1797. It took three years to wind up the estate and decide to sell Clapham Wood Hall, and in April 1800 he returned to the place of his birth (the only recorded instance) to sign the documents to receive his ninth share of the proceeds of £380, or just over £42. This sum, easily the annual pay of a cook, must have helped him enormously in the years immediately following. In 1801, the price of bread peaked because of shortages occasioned by the war with France (then in its eighth year), causing general hardship. In 1802, the family was expanded with the addition of a second daughter, Margaret, born on 17 November, and Faraday seems to have helped with caring for her.

In his mid-forties, James Faraday started to suffer from ill health. Letters that he wrote to a brother in 1807 and 1809 refer to his being in considerable pain. In the latter year, the family moved to 18 Weymouth Street, about half a mile from Jacob's Mews. There James died on 30 October 1810 and was interred six days later (at a cost of £1.10s) in the great dissenting burial ground of Bunhill Fields. Faraday rarely referred to his father in later life, and indeed seems to have known little of his family origins. John Tyndall, a believer in genetic inheritance of intelligence, once asked him if his parents had shown any signs of cleverness, but was rebuffed with Faraday saying that 'He could remember none'. However, on one occasion in the mid-1850s Matthew Noble, who was sculpting Faraday, rattled his chisels and found him 'distrait'; a concerned Noble asked the cause and Faraday replied 'the noise reminded me of my father's anvil, and took me back to my boyhood'.

Faraday's mother lived in Weymouth Street for the remainder of her life, taking in lodgers, and dying there in 1838. We do have some letters from Faraday to her which are affectionate, but, like all sons, concealed things he thought might make her anxious. What all this suggests is that even at this early age Faraday had decided to conceal his feelings. In this respect, he was in line with

2. Bust of Faraday, by Matthew Noble, 1854

the changing sensibility of the time from the robustness of the 18th and early 19th centuries to Victorian respectability, a transition so well captured in the contemporary novels of Jane Austen.

Faraday would have appreciated that without his parents' move from Westmorland in the 1780s, he would never have pursued a scientific career in London. There was one further crucial thing that his parents bequeathed him – the opportunity to become a member of the Sandemanian Church.

Glasites and Sandemanians

Dissent from the state-established Anglican Church had entered
the Faraday family by the early 18th century. Faraday's great
grandfather Richard Faraday, from Keasden (about two miles
south of Clapham), a stonemason and tiler, was a 'separatist'. In
1707, he married Janet Walbanke and they had eleven children, of
whom the tenth was Robert Faraday. To understand how the
Faraday family moved from the rather catch-all term of 'separatist'
into a well-defined, if sometimes porous, group which became
known as the Sandemanians, we must briefly consider the complex
religious history of 18th-century Scotland and northern England.

Following the 1707 Act of Union, which made England and
Scotland a single political entity, the government of the new United
Kingdom of Great Britain met considerable resistance when
attempting to impose English type control on the Presbyterian
Church of Scotland. During the 1720s, John Glas, the minister of
Tealing Church near Dundee, preached against state involvement
in the church, citing John 18:36 'My kingdom is not of this world'.
In 1730, he was expelled from the Church and shortly afterwards
moved to Dundee, where he established his first independent
chapel, later founding others in Edinburgh and Perth. Glas was not
alone in holding dissenting views, and a few who were similarly
minded independently founded groups joined in communion with
the congregations he established. In Edinburgh, Glas met and
converted to his views Robert Sandeman, a student at the
University. Sandeman, who in 1737 married Glas's eldest daughter
Katharine, became an Elder of the Perth Church in 1744. Following
his wife's death in 1746, Sandeman spent 11 years studying the Bible
which resulted in his book *Letters on Theron and Aspasio* (1757),
read two years later by the Yorkshire evangelist Benjamin Ingham.

Ingham, a follower of John Wesley at Oxford University, had
struck out on his own and by the early 1740s had established 60

congregations in Yorkshire. One of these was near Clapham and the register of its Inghamite church lists Robert Faraday and his wife-to-be, Elizabeth Dean, as members by 1756. Their first three children, including Faraday's father, were baptized between 1757 and 1761 in the Inghamite church, but none of their seven later children would be.

Ingham not only read Sandeman's book, but also Glas's *Testimony of the King of Martyrs Concerning His Kingdom* (1729). He realized that their theological views were similar, and the following year Ingham sent two of his ministers, James Allen and William Batty, to Scotland to obtain more information about the Glasites. What Allen learned, especially about church government, converted him to Glas's position and against that of Ingham. On his return to Yorkshire, Allen precipitated a schism in the Inghamite church – 'that horrid blast from the north' as a contemporary described it. The congregations that joined Allen, mainly along the western boundary of Yorkshire, created the first Glasite churches in England. Despite this evangelical work, Glas and Allen had fallen out by 1764 over an issue affecting the Newcastle church probably founded independently of Allen. Many of the congregations that Allen had established remained loyal to Glas, including that at Clapham, which met at Wenning Bank close to Clapham Wood Hall. So the Faraday family became connected with the Glasite church, and this explains why Robert and Elizabeth Faraday's seven other children were not baptized into the Inghamites. Three of their sons joined the Clapham church, including Richard (presumably before moving to Kirkby Stephen), but James did not join the Glasites at Clapham.

Between 1761 and 1764, Sandeman founded and served the Sandemanian church in London before going to the North American colonies, where he died at Danbury, Connecticut. The London Sandemanian Church was formally established at Glovers Hall on 23 March 1762. It subsequently moved first to Mouth Street and in 1785 to Paul's Alley in the Barbican, where it is marked on an early 19th-century London map as the 'Scotch Church'. The congregation of the Sandemanian Church was divided between those who had made their Confession of Faith and therefore took part in the "Love Feast" (a re-enactment of the Last Supper) and those who merely attended the services. These services were well attended and this doubtless justified maintaining the 200-seat building in Paul's Alley. The number who made their Confessions was relatively small, and it is only they who are recorded in the rolls of the Church; these numbers started declining immediately after Sandeman left for North America and continued to do so until the Church closed in the 1970s.

There is some evidence which suggests that James Faraday's move to London was facilitated by the Sandemanian connection, especially as his employer, James Boyd, belonged to the Church. But James Faraday had yet to make his Confession of Faith and this highlights the point that the Glasite/Sandemanian Church was not just a community of shared beliefs and practices, but was also very much a family connection. Of the 252 individuals whose Confessions of Faith were recorded in London during the 19th century, just over half shared 19 surnames (out of 96) and this does not take account of children of Sandemanian women who took, on marriage, a surname different to those 19. Nearly one-quarter had just five surnames: members of the Barnard, Vincent, Boosey, Leighton, and Whitelaw families were to figure greatly in Faraday's life. While it was possible for someone not born to a Sandemanian family to become a member of the Church, such outside conversions were unusual, partly because the Church did not proselytize (it was up to God to bring people to the Church not the individuals themselves).

Whether the Anglican baptism of James Faraday's first two children suggests that at this point he had turned his back on his family's religious background (an interpretation which the move to London might support), or if there was some other reason (for example, wanting a place for his children in a church school), is not known. But between the births of his second and third children, James Faraday made his Confession of Faith in the London Sandemanian Church on 20 February 1791. This explains why neither Michael nor his younger sister Margaret received Anglican baptism.

By joining the London Sandemanians and thus separating themselves from the state church, James Faraday set his entire family on a course that would affect the rest of their lives. Faraday made his Confession of Faith on 15 July 1821, a month after marrying Sarah Barnard (who had already made her Confession), the daughter of a Sandemanian silversmith. They married in her parish church, St Faith in the Virgin near St Paul's Cathedral, and his banns were read at St George's Hanover Square. Prior to the civil registration of births, marriages, and deaths begun in 1837, all marriages in England, except for Jews and Quakers, had to be performed by an Anglican minister – something that injured the consciences of all dissenters.

Sarah is a slightly obscure figure, and although nine years younger than her husband, an 1850s photograph (Figure 3) shows her looking considerably older. Faraday wrote in one of his many affectionate letters to her that she was 'a pillow to my mind'. She was more worldly, managed their domestic affairs (including the servants), and although Faraday and Sarah had no children, their siblings produced a total of 81 nephews and nieces, two of whom lived with them for extended periods.

Sandemanians sought to live a life which they believed was in accordance with the practices of the Early Church, especially as recorded in the Acts of the Apostles. This described two types of office holders, Deacons and Elders. The task of the Deacon was to help with pastoral duties such as visiting the sick. Faraday held this office from 1832, roughly at the time when he began supporting various orphanages. In 1840, he was elected an Elder. His duties included preaching, baptizing infants, and presiding at the Love

3. Michael and Sarah Faraday, c. 1850s

Feast, both in London and at Old Buckenham in Norfolk, where there was a small Sandemanian community. Those non-Sandemanians who went to hear Faraday preach, because of his reputation as a lecturer, were disappointed – 'There was no eloquence.'

Old. Buckenham

1862. August. 31. — This is to certify that Alice, Louisa, daughter of Elisha and Susanna Loveday, born the 4th august 1862, was this day baptised by me

M Faraday.

In the presence of me Jane Barnard

X

4. **Baptism certificate of Alice Louisa Loveday signed by Faraday**

Because of its schismatic origin, the Sandemanian Church, miniscule as it was, was prone to division. A major schism occurred at the end of March 1844 when Faraday and 13 others were excluded from the Church, the most serious event that could befall a Sandemanian. Just what caused this is unclear, but Faraday was restored several weeks later, together with most, but not all, of the others. According to the sect's rules, a second exclusion would be permanent, and ironically in 1850 Faraday nearly brought this on himself by doubting the biblical basis of this punishment, but drew back in time and accepted Church discipline. In 1860, Faraday was again elected to the office of Elder which he held until he laid it down in 1864 during a general reduction of his workload. He died on 25 August 1867; there was no attempt to bury him in Westminster Abbey as would happen to the naturalist Charles Darwin 15 years later. Five days after his death, Faraday's cortège stopped briefly at the Royal Institution before moving to Highgate Cemetery where he was buried, with 'no ceremony', in an unconsecrated plot purchased by his brother-in-law, George Barnard, and close to other Sandemanian graves. In death, as in life, Faraday rejected the presumption of the state church to represent the entire nation.

Two instances illustrate the social effect of Faraday's opposition to the state church. Until well into the 19th century, neither Oxford nor Cambridge University admitted non-Anglicans to their degrees (this was one reason why University College, London, was founded in the late 1820s without religious tests). When Oxford conferred an honorary doctorate on Faraday and three other dissenters in 1832, during the British Association's second meeting, members of what shortly became the Tractarian Anglo-Catholic movement, such as John Henry Newman, vocally condemned both dissent and the Association in the context of their opposition to the reform of the House of Commons that year. Another example of the effect of being a Sandemanian was that Faraday would never step foot inside an Anglican church. Thus he declined invitations to attend the Duke of Wellington's funeral in St Paul's Cathedral (1852) and the Princess Royal's marriage in the Chapel Royal (1860), events for which invitations were much sought.

All this is not to say that Sandemanians were a Puritan sect in the way conventionally understood. For instance, Faraday enjoyed theatre and opera, would drink (in moderation), but disliked smoking, would not gamble (since in the Sandemanian view the casting of lots was biblically forbidden) and did not have affairs (although some women, most notably Ada Lovelace, were interested). What made the sect startlingly different from others was their tolerant approach to others. Once Nicholas Wiseman (Cardinal Archbishop of Westminster, 1850–65) asked Faraday 'if, in his deepest conviction, he believed all the Church of Christ, holy, catholic, and apostolic, was shut up in the little sect in which he bore rule. "Oh no!" was the reply; "but I do believe from the bottom of my soul that Christ is with us"'. On another occasion, one Sunday morning on his way to church, he met a fellow electrician Alfred Smee and his daughter going to St Paul's Cathedral on which he commented 'we are then all three bound for the one great object'. Faraday formed friendships amongst the scientific community with colleagues of differing religious views. One of the closest friends of his youth, Benjamin Abbott, belonged to the Quakers. He was close to the geologist and Dean of Westminster, William Buckland, as well as William Whewell, later Master of Trinity College Cambridge – both Anglicans. It was the institutions of orthodox religion that Faraday rejected, not the individuals who adhered to them.

For Faraday, church, family, and work and were all intimately related. His life and work can only be understood in terms of his religious beliefs and practices. All else, his science, his family life, his charitable work, his involvement with the state and its agencies, and so on, followed from these beliefs. Although each individual in the 19th century had a history of faith or its loss, so far as the evidence allows us to tell, Faraday's religious beliefs remained constant after his Confession of Faith. But this was not made until he was nearly 30, and we must now consider the influences on him before then and especially during his apprenticeship.

George Riebau

In the early 19th century, boys of Faraday's social class began work in their early teens. Shortly after his 13th birthday, on 22 September 1804, Faraday was employed as a newspaper-cum-errand boy by George Riebau who ran a bookshop and stationers at 2 Blandford Street very close to Jacob's Mews. Listed in the London directories from the 1780s onwards, Riebau worked originally at 439 Strand.

Like many booksellers during the 1790s, Riebau became associated with radical politics, belonging to the London Corresponding

Society. More interestingly, he published the works of the millenarian Richard Brothers. Millenarians, who flourished during the Civil Wars of the mid-17th century, reappeared during the wars with France that commenced in 1793. They believed, following the Revelation of St John the Divine 20:4, that the Second Coming of Christ was imminent and would usher in a thousand years of His rule. That such beliefs commanded attention indicates the anxiety which many felt as to the outcome of the war with France.

Brothers described himself as 'King of the Hebrews' and suggested that George III should yield his crown to him. Riebau published many of Brothers's prophetic writings and those of his followers, and one imprint stated 'Printed for George Riebau, bookseller to the King of the Hebrews'. Brothers was arrested on 4 March 1795 and detained in a lunatic asylum for the following 11 years. One of his followers in the House of Commons raised Brothers's plight there, but his motions were never seconded. Nevertheless, Brothers retained some support, including Riebau's, who continued to publish his writings, written in the asylum, and those of his followers, until at least 1805, by which time he had moved to Blandford Street, where he remained until 1836.

As Riebau's activities suggest, the term 'bookshop' had rather broader connotations than now. His business covered the publishing of books, selling and binding them, as well as selling and lending newspapers and journals. It was in delivering (and if necessary collecting) journals and newspapers for Riebau that Faraday began his working life. According to a niece, in later life when Faraday saw a newspaper boy he would pass a complimentary comment and reportedly once said 'I always feel a tenderness for those boys, because I once carried newspapers myself'.

Faraday was clearly successful at carrying out the tasks given him by Riebau for on 7 October 1805, just after his 14th birthday and precisely two weeks before the Battle of Trafalgar, he was apprenticed to Riebau for seven years to learn 'the Art of Bookbinding Stationary and Bookselling' as the indenture stated. Furthermore, 'in consideration of his faithful service no premium is given' – a significant saving for his father. Faraday moved to Blandford Street, where Riebau housed and fed him. In exchange, Faraday promised to serve him faithfully and, as that standard indenture form stated, would not commit fornication or haunt taverns or playhouses, etc.

According to an 1809 letter of his father's, Faraday had 'a very good master and mistress, and likes his place well. He had a hard time for some while at first going; but, as the old saying goes, he has rather got the head above water, as there is two other boys under him'. This sounds like a conventional experience of an apprentice, especially the reference to having a hard time at first. Nevertheless, Faraday learned his chosen craft well, and a number of books bound by him survive.

Quite what Faraday made of Riebau's millenarian connections is not known, but as an apprentice at Riebau's he doubtless knew of them. It would seem that this was not his only connection with millennialism. According to one of his nephews, Frank Barnard, Faraday 'in his younger days...had his period of hesitation, of [religious] questioning'. Barnard also said that he had heard that Faraday had visited Joanna Southcott 'perhaps to learn what that women's pretensions were'. Originally from Devonshire, where she had declared that she would bring forth the second Christ, she moved to Paddington in 1802. In London, she gained a wide following and it would perhaps not be too surprising if Faraday had visited her; whatever his commitments at this time, he later treated Revelation with the same regard as the rest of the Bible.

Such doubts and such a visit would mesh well with the idea that in his teens Faraday was seeking some means of finding secure knowledge. This interpretation is supported by an 1858 letter to Auguste De La Rive where he recalled:

> I was a very lively, imaginative person, and could believe in the Arabian nights as easily as in the Encyclopaedia [Britannica]. But facts were important to me & saved me.

What saved Faraday, and what was certainly not conventional about his apprenticeship, were the scientific interests he developed by reading 'in the hours after work' many of the books that he bound. By 1809, Faraday was keeping a 'Philosophical Miscellany' in which he recorded the impressive breadth of his scientific reading. His reading included the scientific entries from the *Encyclopaedia Britannica* which especially helped him at the beginning of his philosophical studies. Unknown (presumably) to him, was that most of these entries, including that on electricity, were written by James Tytler, who for a while was closely connected with the Edinburgh Glasite Church.

Other books of particular value to him included *The Improvement of the Mind* by the dissenting minister and hymn-writer Isaac Watts and *Conversations in Chemistry* by the Anglo-Swiss author Jane Marcet. From Watts's book, first published in 1741 with many subsequent editions, Faraday gained advice about how best to obtain knowledge, which included the skills of writing letters, keeping a common-place book, and organizing a small essay circle. Marcet's *Conversations*, published anonymously in 1806 (the fourth edition, in 1813, first named her), made such a profound impression on him that following her death, Faraday commented on its importance during his apprenticeship:

> I could trust a fact and always cross-examined an assertion. So when I questioned Mrs. Marcet's book by such little experiments as I could find means to perform, and found it true to the facts as I could

understand them, I felt that I had got hold of an anchor in chemical knowledge, and clung fast to it.

The number of topics from *Conversations* noted in his 'Philosophical Miscellany' supports this memory, and a friend's later recollection suggests Riebau allowed Faraday to perform experiments in his shop and indeed that he had sufficient apparatus, including electric batteries, to constitute a 'miniature Laboratory'. What is not so easy to confirm was whether Faraday's methodological prescriptions stemming from Marcet were already in place as early as 1809.

Faraday retained pleasant memories of his time as an apprentice bookbinder. One day in late 1855, on his way to Baker Street with Tyndall, they visited Blandford Street. The shop was still a stationers and, according to Tyndall, Faraday became highly animated as he showed him his old workplace. However, Faraday's memory, as he knew himself, played all sorts of tricks with him, and against these pieces of evidence one has to note that his contact with Riebau after 1815 was limited, and he later referred to trade as 'vicious and selfish' which he gave as the reason for wishing instead 'to enter into the service of Science'.

His apprenticeship ended on 6 October 1812, and he started his journeyman career with Henri De La Roche, a bookbinder of 5 King Street, a westward continuation of Blandford Street. This geographical proximity suggests that Riebau helped secure the position, something which Faraday hints at in a letter. Faraday moved from Blandford Street to live with his widowed mother in Weymouth Street. De La Roche has been described as a passionate French émigré who gave his assistant 'much trouble', but offered to leave Faraday his business if he would stay with him. Riebau noted that Faraday 'received a Guinea & half [£1.11s.6d] per week which I think very fair wages for a Young man just out of his time'. Shortly after his appointment, Faraday complained about the constraints on his time that employment brought and wished 'to leave at the first convenient opportunity'. His memory of his employer improved with time, and a couple of years later he wrote that he was 'grateful for the goodness of Mr. de la Roche'; nevertheless, he sought a career in science.

Chapter 2
A career in science

John Tatum

Faraday's apparently conscious decision to pursue a scientific career was a strange one. The number of scientific practitioners in Britain was small, with most earning their living in other professions or possessing private wealth. In 1812, Anglican clergymen (including eight bishops) made up 11% of the Royal Society's 570 Fellows, while one-quarter had hereditary titles, and one-fifth were MPs. With just over 100 men in paid scientific posts, science was frequently pursued by Anglican clergymen (Buckland or John Barlow) or army officers (Edward Sabine or Roderick Murchison) or naval officers (Robert Fitzroy or Francis Beaufort). Furthermore, many of those who practised science had some sort of medical background, and these comprised 17.5% of the Royal Society. Of those with whom Faraday would work, Humphry Davy and William Brande trained as apothecaries, while Thomas Young and William Wollaston were both medical doctors. For Faraday, who needed to earn his living, the decision to leave a safe occupation and pursue a scientific career without one of these appropriate backgrounds must have taken enormous courage, a belief in his own abilities and also, perhaps, in the protection of providence.

There was only so much that Faraday could learn from books and the few experiments he performed in Riebau's shop. Fortunately for him, there were many lecturers in London providing courses on scientific subjects. It was those delivered by the silversmith John Tatum which soon exerted a major influence on Faraday's developing interest in science. Tatum delivered, normally on a Monday evening, courses of scientific lectures at his house, 53 Dorset Street (now Dorset Rise just to the north-west of Blackfriars Bridge). These lectures, advertised by handbills in the streets and shop windows, were open to both ladies and gentlemen on the payment of one shilling per lecture. Faraday later noted that he 'attended twelve or thirteen lectures between 19 February 1810, and 26 September 1811', that he was 'allowed' by Riebau to attend the lectures (as required by the terms of his indentures) and that his brother, Robert, gave him the money to attend some of them. Faraday's four volumes of notes record his attendance at thirteen lectures, of which seven were on electrical subjects. At Tatum's lectures and the associated City Philosophical Society, Faraday met a number of like-minded people with whom he formed life-long friendships. These included Benjamin Abbott, who later became a teacher; John Huxtable, who lent Faraday various chemical textbooks and became an apothecary; Edward Magrath, who later ran the Athenaeum; and Richard Phillips, who became a distinguished chemist.

Faraday's diligence in pursuing his study of science was rewarded in early 1812 when Riebau showed Faraday's notes of Tatum's lectures to one of his customers. This particular customer was the son of William Dance, who lived nearby in Manchester Square. The following day Riebau, at Dance's request, showed him the notebooks. So impressed was Dance with them that he gave Faraday tickets to attend the last four lectures to be delivered by Humphry Davy, Professor of Chemistry at the Royal Institution, of which Dance had been a Proprietor since 1805.

The Royal Institution

The Royal Institution of Great Britain was founded formally at a meeting held on 7 March 1799 at the Soho Square house of the

President of the Royal Society, Joseph Banks. At that meeting, 58 men agreed to form an 'Institution For diffusing the Knowledge, and facilitating the general Introduction, of Useful Mechanical Inventions and Improvements; and for teaching, by Courses of Philosophical Lectures and Experiments, the application of Science to the common Purposes of Life', and each contributed the substantial sum of 50 guineas to become the founding Proprietors of what was initially called the 'Institution' in deliberate imitation of the famous Istituto delle Scienze e delle Arti in Bologna. Thus was founded what became and remains one of Britain's key scientific institutions, which formed the model for others elsewhere in the country and overseas, most notably the Smithsonian Institution in Washington, James Smithson being an early Proprietor of the Royal Institution.

The men who came together to form the Royal Institution had known one another during the previous decade in a number of overlapping contexts including membership of the Society for Bettering the Condition of the Poor, the Board of Agriculture, and connections with the East India Company. Many were also landowners and aristocrats. In addition, a significant number of them were associated with the Anglican evangelical revival. Thus an interest in the way science could aid philanthropy, agricultural improvement, industrial processes, and imperial expansion, at a time of the war against France and when the social effects of major industrialization were beginning to be felt, provided strong motivations to found the Royal Institution.

The two tasks to which the early Royal Institution gave priority were to gather more Proprietors and to acquire a building in which to deliver its programme of lectures and advice. On the market because its former owner, John Mellish, had been killed by a highwayman on Hounslow Heath, 21 Albemarle Street (off Piccadilly) was seen right from the start as a potential home for the new institution. Purchased in mid-1799 for £4,850, the Royal Institution remains there. Because this was a gentleman's town house, built in stages during the 18th century, considerable construction was required to convert it into a scientific institution with lecture theatres, laboratories, libraries, and display areas. The architects John Soane and Henry Holland (both Proprietors) provided advice on the conversion which, under the supervision of Benjamin Thompson (until 1802 when he departed abruptly for Paris), was executed by the clerk of works Thomas Webster. A temporary lecture room was quickly created, and to Webster's design the semi-circular two-tiered lecture theatre, which could hold more than 1,000 people, was soon erected at the building's northern end.

Doubtless to demonstrate the Institution's usefulness and thus attract more Proprietors and Subscribers to help pay for the construction work, the lecture programme was established quickly. On 11 March 1800, just over a year after its founding, the chemist Thomas Garnett delivered the first lecture in the Royal Institution. To further aid recruitment, George Finch, Earl of Winchilsea and Lord of the Bedchamber to George III, was elected the first President in June 1799. Using his influence, he persuaded the King to be Patron and thus provide the Royal handle to the Institution.

However, all was not well. Problems in cashflow led to Garnett not being paid on time and he resigned, to be replaced by Thomas Young, who was not an inspiring lecturer and lasted only one season. The funding crisis, which culminated in 1810, resulted in a major reordering of the Royal Institution's governance and administration. The Proprietors were persuaded to relinquish their ownership rights in the Institution and to become Life Members; Subscribers either paid an annual subscription as a Member or compounded as Life Members; all this ensured that there would be a basic income-stream from membership. The Members elected a committee of Managers to run the Royal Institution, and a committee of Visitors who acted as the financial auditors but also commented on the Institution's activities and the fabric of the building. One of the main architects of this reform was Humphry Davy, whose appointment, when he was aged 22, Thompson recommended in February 1801, and who was promoted to Professor of Chemistry the following year.

Humphry Davy

Born in 1778 in Penzance, the son of a Cornish woodcarver who died in 1794 leaving debts of £1,500, Davy attended Truro Grammar School before being apprenticed as an apothecary in 1795. Like Faraday a decade later, Davy took a strong interest in science during his apprenticeship, especially the chemistry of Antoine Lavoisier. He attracted the patronage of Davies Giddy (Gilbert from 1817), Deputy Lieutenant of Cornwall. At Oxford University, Giddy had been a close friend of Thomas Beddoes, Reader in Chemistry. In 1793, Beddoes left Oxford for Bristol following serious disagreements with the university authorities over the French Revolution which he supported. Towards the end of the decade, Beddoes, with the financial support of the Midland industrialists Josiah Wedgwood and James Watt, both of whose sons suffered from tuberculosis, established a clinic, the Pneumatic Institute, to develop therapies for such diseases using the gases recently discovered by the radical chemist Joseph Priestley. For this project, Beddoes needed an assistant. Giddy got Davy released from his apprenticeship, and in October 1798 he travelled to Bristol to assist Beddoes.

Although he lived in Bristol for just under two and a half years, this was a key period for Davy. He nearly killed himself while experimenting on carbon monoxide, but found inhaling nitrous oxide produced pleasurable effects and consequently was popularly called laughing gas. Once, when under the influence of seven quarts, he wrote in his notebook in one-inch letters, 'Newton & Davy', an indication, perhaps unconscious, of his vaunting ambition to imitate Isaac Newton's achievements a century earlier. Davy also formed close friendships with political radicals (opposed to Pitt's government, perhaps the most repressive since the 17th century) then living in Bristol, such as the publisher Joseph Cottle, the poet Robert Southey, and the poet and philosopher Samuel Taylor Coleridge, to whom he was particularly close. Davy persuaded Coleridge, Southey, and others to self-experiment with nitrous oxide and he published their accounts of the experience in his first book, *Researches, Chemical and Philosophical, Chiefly Concerning Nitrous Oxide* (1800). His friendship with Coleridge resulted in Davy being given the responsibility for seeing through Cottle's press that seminal text of English Romanticism, the second edition of William Wordsworth's *Lyrical Ballads* (1800). This connection with the key English Romantics has always given Davy a unique position in English science – Coleridge reportedly commented that 'Had not Davy been the first chemist, he, probably, would have been the first poet of his age'. Davy's critical facilities, never particularly strong, were further weakened by association and praise from these self-centred Romantics, so that he came to believe that he could do anything and, despite his humble origins, treated with utmost contempt those whom he regarded as socially or intellectually inferior.

Davy moved to London in March 1801 and the Royal Institution fell increasingly under his agenda. He was an enormously attractive lecturer who filled the lecture theatre with his spectacular, not to say dangerous, demonstrations of chemical experiments, famously caricatured by James Gillray in his depiction of the administration of laughing gas to the Royal Institution's Treasurer John Hippisley. But it was not just Davy's lectures that attracted large audiences to the Royal Institution in its early years. Figures such as Coleridge (at Davy's invitation), the musician Samuel Wesley, the chemical atomist John Dalton, and the author and wit Sydney Smith all attracted audiences, and Smith's lectures on moral philosophy were possibly even more popular than Davy's.

The requirement to stage spectacular chemical experiments led the Royal Institution to quickly acquire perhaps the best equipped laboratory in England and one of the best in Europe. In it, Davy used the electric battery (invented by Alessandro Volta in the 1790s) to isolate for the first time a number of chemical elements, including sodium and potassium, and to develop the first coherent theory of electro-chemical action. Davy thus initiated the Royal Institution's involvement in original scientific research, something which its founders had never envisaged.

But lectures and research, no matter how important, were not sufficient to secure the future of the Royal Institution, and Davy continued with its utilitarian programme by, for instance, providing lectures to the Board of Agriculture and researching into leather tanning. In this latter project, he concluded that the processes already used by tanners could not be improved by scientific understanding. This failure of science to improve a practical process highlighted a key problem of the Royal Institution's agenda; despite its rhetoric, it was very difficult to use scientific knowledge and methods for improving technical and agricultural processes. Such disjunctions contributed to the necessary reordering of the Royal Institution in 1810.

Davy's career at the Royal Institution culminated with his being knighted by the Prince Regent (later George IV) on 8 April 1812 and three days later marrying a very wealthy widow, Jane Apreece, who had attended his lectures. A relation of Walter Scott's, she had originally married Shuckburgh Apreece, but he died before inheriting his baronetcy. When they married, with Davy already knighted, she became Lady Davy. Because of her wealth, Davy could retire from being the Royal Institution's Professor of Chemistry at the age of 34. It was a marriage that could easily have been portrayed in one of Austen's contemporary novels: he got the money and she got a title. Anxious to retain their connection with arguably the most famous English chemist of the time, the Royal Institution's Managers appointed Davy Honorary Professor of Chemistry and, continuing as Director of the Laboratory, he retained considerable influence in the Royal Institution after his marriage.

Faraday and the Royal Institution

The lectures delivered by Davy during March and April 1812 were thus his final performances in the Royal Institution's theatre. According to one account, Faraday (possibly accompanying Dance) sat in the gallery behind the clock. In these lectures, Davy dealt with the nature and definition of acidity, a problem at the cutting edge of chemical knowledge. French chemists, following Lavoisier, argued that all acids must contain oxygen (which means 'acid producer') and that therefore muriatic acid contained oxygen. Davy showed that muriatic acid gas was a chemical element which he named chlorine and that muriatic acid was a compound of hydrogen and chlorine (hydrochloric acid, HCl) containing no oxygen. As with Tatum's lectures, Faraday took detailed notes of what Davy said, and the experience must have encouraged him to continue his pursuit of a scientific career.

In July, he told his friend Abbott that he had had the prospect of a position with a large annual salary of £500–£800 which required knowledge of, amongst other things, mathematics and mechanics, which he did not possess: 'Alas Alas Inability'. He also wrote to Banks in 1812 asking to be 'engaged in scientific occupation, even though of the lowest kind'; the reply was 'no answer'. According to Riebau, the journeyman Faraday would occasionally call on him (Blandford Street was on his way home). On one of these occasions, Faraday expressed a wish to meet Davy. Riebau suggested that he write directly to Davy and send his lecture notes and drawings; according to Faraday's account, this suggestion was made by Dance. Whoever made it, Faraday in late December wrote to Davy sending his notes and later recollected that 'The reply was immediate, kind, and favourable'. Dated Christmas Eve 1812, Davy replied that he was pleased with the proof Faraday had given him of his 'great zeal power of memory & attention' and would be happy to see him at the end of January following his return to town.

Davy at this time was suffering from injuries sustained following an explosion caused by combining nitrogen and chlorine. Glass had penetrated his eyes, his sight badly damaged in consequence, and although optimistic in a letter to his brother, John (an army physician), in the middle of November, two months later his eyes were still inflamed. It was not until early April that he declared himself 'quite recovered'. Because of his eyes, he had to employ amanuensises, and both Faraday and Riebau recorded that Davy initially employed Faraday in this capacity for a few days, although no contemporary document in Faraday's hand has been found. The story suggests that Davy interviewed Faraday after his return to London some time in January 1813, and it was probably on this occasion that he advised him to 'Attend to the book binding'. One presumes that De La Roche allowed Faraday the time off to be Davy's amanuensis and this doubtless explains Faraday's view of the goodness of his employer.

On Friday 19 February 1813, William Harris, the Superintendent of the Royal Institution, heard a 'great noise' in the lecture theatre. On investigation, he found that the laboratory assistant William Payne and John Newman, instrument-maker to the Royal Institution, 'at high words'. Newman charged Payne with neglecting his duty in not attending on the new Professor of Chemistry, William Brande, and complained to Harris that Payne had struck him. This charge was considered by the Managers at their next meeting three days later and they promptly sacked Payne; there was no mention of finding a replacement. However, at their following meeting, on 1 March 1813, it was recorded that:

Sir Humphry Davy has the honor to inform the Managers that he has found a person who is desirous to occupy the situation in the Institution lately filled by Wm. Payne[.] His name is Michael

Faraday[.] He is a youth of 22 years of age; As far as Sir H. Davy has
been able to observe or ascertain he appears well fitted for the
situation[.] His habits seem good, his disposition active and
cheerful and his manner intelligent[.] He is willing to engage
himself on the same terms as those given to Wm. Payne at the time
of his quitting the Institution.

The terms were £1.5s weekly, a pay reduction, but this was
compensated for by his being given a room in the Royal
Institution to live. Whether Davy was asked to find a replacement
is not clear, but nevertheless his recent contact with Faraday must
have prompted him to see if Faraday was still interested in
scientific employment. According to Abbott, Faraday was
undressing for bed when Davy's footman arrived with a note
requesting Faraday to call the following morning. This Faraday
did, and it was probably on this occasion that Davy told him that
'Science was harsh mistress', poorly paid, and smiled at his
'notion of the superior moral feelings of philosophic men, and
said that he would leave me to the experience of a few years to set
me right on that matter'. Despite these warnings, Faraday
accepted the offer and commenced the new career he had so
actively sought.

Faraday was employed by the Royal Institution rather than by
Davy personally and it is misleading to refer to him as Davy's
assistant. Nevertheless, Davy's prestige within the Royal
Institution meant that he used Faraday's services frequently. Early
in his time there, he helped Davy continue the experiments that
had damaged his eyes. The danger continued, Faraday reporting
'I have been engaged this afternoon in assisting Sr. H. in his
experiments…during which we had two or three unexpected
explosions'. Fortunately, Faraday wore a mask to protect his eyes,
though this did not prevent damage to his hands. Despite these
difficulties, Davy wrote a paper on these experiments for the
Philosophical Transactions, the manuscript of which is in
Faraday's hand.

Europe

Faraday hardly had time to learn the ropes at the Royal Institution before Davy, probably in early September 1813, proposed that he should accompany him as his 'philosophical assistant' on a tour of Europe and Asia projected to last three years. Like most of the Romantics of his time, Davy was drawn to the Continent, where things could be seen and experienced which were simply not possible in Britain. All the Romantics – Wordsworth, Coleridge, Southey, and later Byron, Shelley, and Keats – acquainted themselves thoroughly with mainland Europe. In many ways, their visits were in succession to the Grand Tours of the 18th-century British aristocracy, and many of the Romantics followed a similar itinerary especially in Italy. Only Davy, of that group in Bristol, had never ventured outside the British Isles and must have felt himself terribly provincial. Thus when the opportunity to visit the European mainland presented itself, it is little wonder that he took it with alacrity.

In retrospect, it is quite remarkable that the government of the country which Britain had been fighting more or less continuously for 20 years provided Davy with a passport to visit the French empire. Furthermore, the British government made no objection, although *The Times* hoped that Napoleon would intern Davy in Verdun for the duration. Davy's relations with his French counterparts were difficult though by no means always acrimonious. For example, on 7 December 1807, the Académie des Sciences awarded Davy the Volta prize of 3,000 francs founded by Napoleon for work on galvanism. But six months later, Davy was not elected a corresponding member of the Académie's chemistry section, receiving a derisory six votes. One presumes that Napoleon was induced to provide a passport for Davy to visit France because of the award of the Volta prize.

At 11 o'clock on the morning of Wednesday, 13 October 1813, Davy, Lady Davy (who created some tensions during the tour), her maid, and Faraday left London. Faraday noted in the diary that he kept for the entire journey that previously he had never travelled more than 12 miles from London. Thus at about the time Davy's coach passed Joseph Banks's country house, Spring Grove in Isleworth, on the road leading west from London, Faraday entered upon unknown territory. Sailing from Plymouth, they arrived in Paris on 28 October 1813 on the first leg of a journey that would last 18 months.

Davy was in a difficult position in Paris. He was the guest of a man whom most Englishmen of his class regarded as a tyrant (the fact that worse was to follow in European history does not affect the perception then), yet somehow he had to establish that he was not collaborating with the enemy, to use a somewhat later term. His response was to continue the war by other means, and his reaction to the loot displayed in the Galerie Napoleon (the Louvre) – 'What

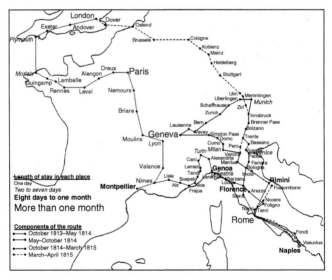

5. Map of Continental tour

an extraordinary collection of fine frames' – was one way of doing this.

In Paris, Davy demonstrated, much to the annoyance of the French chemists, that iodine was a chemical element; further south, in Montpelier, he experimented on the torpedo (an electric fish), and in Florence he demonstrated using the Duke of Tuscany's large burning lens that diamond was a form of carbon. In Naples, they heard of the allied victory over Napoleon which was celebrated by the subjects of the allied powers with a party on Vesuvius described vividly by Faraday. The following year, they were in Naples again and there heard of Napoleon's escape from Elba, whereupon Davy cut short the tour and returned rapidly to England, arriving in London on 23 April 1815.

It is striking how many prominent men of science in the mid-19th century spent considerable time abroad at relatively early points in their careers. Charles Darwin sailing round the world on HMS *Beagle* is the most obvious example. It is difficult to imagine his career without that voyage, as it is for many others such as John Herschel (at the Cape), Alexander Humboldt (South America), Edward Sabine (the Arctic), William Thomson (Paris), or Roderick Murchison (Siberia). One can, however, imagine without too much difficulty that Faraday's career would have been much the same without the Continental tour, and this may explain why he made so few subsequent references to it.

Back in London

For about a month after his return, Faraday was effectively jobless, but in the middle of May he was reappointed to his old position with accommodation. For the remainder of his life, he worked for the Royal Institution, receiving promotions during the 1820s, until in 1833 the wealthy and eccentric patron of the arts and sciences, John Fuller, created and endowed especially for him the Fullerian chair of chemistry which brought Faraday an extra £100 annually, with no additional duties. The Royal Institution was the centre of his scientific career where he formed many of his close associations, for example with the astronomer James South, to whom Faraday gave the electro-static machine he made whilst an apprentice and his notes of Tatum's lectures.

Davy spent much of the summer of 1815 in Scotland. There he received a letter from the Rector of Bishopwearmouth asking, following some disasters in the north-east, if he could find a way of lighting coal mines safely without causing explosions. On his way back to London, Davy stopped in Newcastle to collect some samples of fire-damp (methane, CH_4) for analysis. In an intense two-month period starting in mid-October 1815, Davy, with Faraday's assistance (which he fully acknowledged), invented the gauze miners' safety lamp which proved remarkably successful. During the next two years, Davy fought a ferocious priority dispute with the Newcastle mining engineer George Stephenson who had invented a similar lamp at precisely the same time. Davy eventually defeated Stephenson by dubiously claiming that his lamp was based on scientific principles whilst Stephenson's was not. This claim, together with his elevation to a baronetcy, put Davy in line to become President of the Royal Society following Banks's death in 1820.

It is easy, in retrospect, to see that electing Davy following Banks's 42-year presidency was unwise. William Wollaston, caretaker President for a few months following Banks's death, was more than happy to give way to Davy, suggesting that he at least recognized that serious problems existed. The major issue facing Davy was making the Fellowship of the Royal Society more scientific. This entailed reducing the number of non-scientific Fellows who were elected, and Davy's attempt to do this created considerable animosity from various factions within the Society. Davy's political incompetence meant that the Royal Society was not reformed until 1847, half a generation after most other English institutions.

Davy's problems at the Royal Society would have a direct bearing on Faraday's election to the Fellowship.

During the latter half of the 1810s, Faraday built up his reputation as a chemist by, for example, undertaking some important metallurgical work on steel alloys with the surgical instrument-maker James Stodart. In early 1821, he was commissioned by his old City Philosophical Society friend Richard Phillips to review for the *Annals of Philosophy* the literature on the phenomenon of electro-magnetism discovered by the Danish savant Hans Oersted the year before. To understand the often unclear published descriptions of electro-magnetic phenomena, Faraday repeated many of the experiments. As a result, on 3 and 4 September 1821 he found that he could make a wire carrying an electric current move round a magnet. He called this phenomenon 'electro-magnetic rotations', and it can be viewed as the principle behind the electric motor. His rapid publication of this discovery in the *Quarterly Journal of Science* (which he was editing while Brande was away) got him into some trouble as a rumour went around suggesting that he had plagiarized some of Wollaston's electro-magnetic work. However, Wollaston loftily declined to pursue the matter, though the accusation continued to surface thereafter.

One of these occasions occurred in early 1823, when Faraday liquefied a gas (chlorine) for the first time while doing some experiments at Davy's suggestion. Davy believed that Faraday had not properly acknowledged his contribution and was reported publicly as making from the President's chair some disparaging remarks about Faraday's originality:

> we owe to the sagacity of Dr. Wollaston, the first suggestion of electro-magnetic rotation; and that, had not an experiment on the subject, made by Dr. W. in the laboratory of the Royal Institution, and witnessed by Sir Humphry, failed, merely through an accident

which happened to the apparatus, he would have been the discoverer of that phenomenon.

During April 1823, Phillips, only elected himself the previous year, organized Faraday's nomination certificate to the Fellowship, with Wollaston's name placed prominently at the top. At the end of May, Faraday noted that Davy was angry and asked him to take his form down. Faraday stood his ground and said that only his proposers could do that. 'Then he said, I as President will take it down. I replied that I was sure Sir H. Davy would do what he thought was for the good of the Royal Society' – a somewhat cheeky response, and Davy did not take it down. Davy by now was so desperate that he spent an hour walking round the courtyard of Somerset House with one of Faraday's proposers arguing that Faraday should not be elected. In the end, Faraday was elected a Fellow on 8 January 1824 and admitted, by Davy, to the Society a week later.

In part, Davy's opposition to Faraday's election stemmed from his desire to be seen to break from Banks's system of patronage. If Davy did not publicly oppose Faraday's election, it would be assumed that he wanted his protégé elected – a return to Banksian practices. Faraday felt hurt and later wrote that he was 'by no means in the same relation as to scientific communication with Sir Humphry Davy after I became a Fellow of the Royal Society as before that period'. Nevertheless, Faraday published most of his major work in the Society's *Philosophical Transactions*, and was rewarded by receiving both the Society's Copley and Gold Medals twice – the modern equivalent would be winning the Nobel Prize. Despite such recognition, his experiences with Davy doubtless contributed to his declining the presidency in 1848 and again 10 years later. On that occasion, he commented that if he accepted, he 'would not answer for the integrity of my intellect for a single year'. However, despite his view that the presidency was a corrupting office, Faraday never lost his regard for Davy's science and once chided J. B. A. Dumas for criticizing him.

Chapter 3
Science and practice

Davy's exploitation of Faraday

There is a fine line between patronage (where both sides gain
something from the relationship) and exploitation. It is an open
question when precisely Davy crossed that line so far as Faraday
was concerned. Certainly after Faraday's election to the Royal
Society, Davy sought only to exploit Faraday's undoubted talents.
He involved Faraday, with no regard to his interests, with three
time-consuming projects. These were the founding of the
Athenaeum, the project to protect the copper bottoms of naval
vessels, and the attempt to improve optical glass.

One strategy pursued by Davy to make the Royal Society more
scientific was founding an elite club that those non-scientific
figures who might have previously aspired to the Royal Society
could join instead. Following an exchange of correspondence
between Davy and the conservative author, parliamentarian, and
Secretary to the Admiralty, John Wilson Croker, a meeting of
interested parties was held on 16 February 1824 to establish such a
club, initially called The Society. Faraday was invited to become its
first secretary, and during February and March he sent several
hundred (perhaps more) printed letters to various eminent men of
science, artists, writers, and churchmen inviting them to join. He
acknowledged those who did and in mid-March sent a printed

reminder to those who had not responded listing those who had already joined.

At a meeting on 17 May, it was agreed to change the club's name to the Athenaeum and offer Faraday £100 annually to be secretary. This he declined and suggested that his old City Philosophical Society friend, Edward Magrath, should be appointed. Magrath (who held this position until retirement in 1855) could not take up his new duties immediately, and so Faraday dealt with a rights of light issue at the Athenaeum's temporary home in Waterloo Place until the end of May 1824. Faraday was elected a member with the first year's subscription waived. Later he advised on the ventilation and lighting of the clubhouse (designed by Decimus Burton) as well as supporting the election of new members. In 1851, he decided that he could no longer afford the fee and resigned from membership. He later complained that belonging to the Athenaeum had cost him nearly £200, for which he had just one dinner.

The second example of Davy's exploitation of Faraday arose at the beginning of 1823 when the Navy Board (which provided the Royal Navy's civilian administration) approached Davy about the possibility of protecting the copper sheeting of warships from the corrosive effects of seawater. The naval budget had been reduced by 71.4% since the end of the war in 1815, and hence the Navy Board was seeking to lower expenditure. If the frequency with which ships needed to be dry docked to replace their corroded copper could be reduced, then significant savings would be made.

During 1823, the Navy Board provided Davy with information about copper corrosion and following his return from holiday at the end of October, he began investigating the problem. By mid-January 1824, he concluded that there existed an electrical reaction between the copper and the oxygenated seawater (no corrosion occurred when oxygen was not present) which allowed the formation of various copper salts. Thus, he reasoned, that if the electrical polarity between the copper and the seawater was reversed, the corrosion would cease. In his *Elements of Chemical Philosophy* (1812), he had ranked the electro-chemical reactivities of various metals. Zinc was much more electro-positive than copper which suggested that a relatively small amount attached to the copper would prevent the corrosion.

The day after he personally informed Croker of this method, the Admiralty ordered that practical tests should be carried out on three warships moored in Portsmouth Dockyard. Starting in mid-February 1824, Davy's "protectors" as they were called (doubtless in deliberate imitation of Davy's lamps) were attached to their copper, the state of which was monitored in the ensuing months. Faraday, who undertook most of the follow-up experiments, visited Portsmouth once. At the end of April, satisfied that the tests were successful, the Navy Board drafted an order that the entire fleet be fitted with the protectors. This was issued by the Admiralty on 21 May, and the fitting programme was undertaken during the remainder of the year and into 1825. However, early in 1825 problems began to appear, and by the summer it was clear that the Navy faced a major disaster. Ships returning from the West and East Indies were found to have their bottoms, though preserved, fouled with seaweeds, barnacles, and suchlike. Because of the protectors, no longer were the poisonous salts produced by the corroding copper being released into the water to kill the source of the fouling. Davy, to some degree, recognized that this might be a problem and had tried by varying the ratios of protectors to copper to prevent it, but such was the rush and the inadequacy of the Portsmouth trials, that the full extent of the problem did not become apparent until after the protectors were operational. On 19 July 1825, the Admiralty ordered the removal of the protectors.

There then followed the political task of allocating the blame for the disaster. The Navy Board had protected itself by doing only what the Admiralty ordered. Hence in the eyes of the Admiralty, and Croker in particular, Davy was to blame. This failure doubtless contributed to Davy's ill health and premature resignation as President of the Royal Society on 6 November 1827.

While it still appeared, in early 1824, that Davy had scored a major success with the Admiralty, he proposed, as Chair of the Board of Longitude, forming a joint committee of the Board and the Royal Society for the purpose of improving optical glass. This would be valuable for producing better navigational instruments and was therefore within the Board's remit. The joint committee was established, its membership included Davy, Wollaston, and Gilbert, together with the optician George Dollond and astronomer John Herschel. The committee first met at the end of May 1824 and appointed the glass-making firm of Pellatt and Green (whose premises were south of Blackfriars Bridge) and asked Faraday to analyse chemically the glass produced. The project did not go well, and a year later a subcommittee was formed comprising Herschel, Dollond, and Faraday. Faraday was to supervise making the glass, Dollond was to grind it, and Herschel was to determine its optical properties. Again, there was no success, and by this time the disaster of the copper protection was apparent, and Davy last chaired the joint committee in May 1826. There was then a hiatus while the joint committee was reorganized with first Wollaston and then Gilbert taking the chair.

It took another year before it was decided to build a glass furnace in the Royal Institution and for Faraday to make the glass himself. High-quality optical glass needs homogeneity and a high refractive index, which can be obtained only by doping the glass with a heavy metal such as lead. As an ingot cools, the metal sinks to the bottom and thus has to be stirred, which introduces bubbles and striations. It was known that these problems could be overcome because the Bavarian optician Joseph Fraunhofer had succeeded a decade earlier. He kept his processes secret and Faraday's task was basically to replicate Fraunhofer's methods. In December 1827, he began two years of arduous and ultimately fruitless work spending nearly two-thirds of his available working time on the project, in which he recorded making 215 ingots of glass. He was provided with an assistant, Charles Anderson, formerly a sergeant in the Royal Artillery. He proved ideal for Faraday who retained him at the Royal Institution for the remainder of their working lives.

Faraday, unable to achieve success, became frustrated with the amount of time consumed by the glass project, believing it was preventing him from making 'three or four philosophical discoveries'. So desperate was he that in May 1829 he opened negotiations with the Royal Military Academy, Woolwich, to move there as Professor of Chemistry. As it happened, Davy died in Geneva at the end of May, and during the next few months the circumstances which had involved Faraday in the glass project quickly dissolved and he was able to abandon it. Nevertheless, not wishing to be so exploited again, Faraday made himself partially financially independent of the Royal Institution by being appointed from December 1829 part-time Professor of Chemistry at Woolwich, receiving £200 annually for providing 25 lectures. From then until 1851, he spent two days a week at the Academy during their terms, which, whilst a considerable commitment of his time (his 'only estate'), was, nevertheless, an enormous improvement compared to that devoted to making glass. For two decades, cadets of the Royal Artillery and the Royal Engineers learned their chemistry from Faraday, though exactly how useful this knowledge was to them in their later careers is not clear.

Faraday, science, and the state

Despite the problems that Davy had created, and which he had observed at close quarters, Faraday was seen by politicians and civil servants as inheriting Davy's mantle when it came to providing advice for the state and its agencies. Indeed, from the 1830s to the 1860s, one can almost hear the cry "send for Faraday" go up in Westminster and Whitehall when scientific advice was required. Even the state's own experts acknowledged Faraday's value; the Astronomer Royal, George Airy, put it: 'We trouble you as a universal referee or character-counsel on all matters of science'. And Faraday was fully committed to helping the state use science; what he disliked intensely was wasting his time on what he regarded as useless projects, such as the glass work, or tasks that could easily be undertaken by other chemists, the numbers of whom increased with the expansion of the scientific community during his lifetime.

The government paved the way for Faraday's involvement, and at the same time took swift revenge on Davy for the failure of both the copper protectors and the glass work, by abolishing, in 1828, the Board of Longitude, the only state body providing public funds for science. In its stead, the Admiralty established from the beginning of 1829 what they called the Resident Scientific Committee. The proceedings of this Committee immediately followed in the Board's minute book which suggests an institutional continuity, as does the appointment of the Board's former Secretary, Thomas Young, to the Committee. Croker also appointed Edward Sabine and Faraday to the Committee. Young soon died and Sabine was posted to Ireland which left Faraday as the sole scientific adviser to the Admiralty, a role he fulfilled well into the 1850s.

Thus in the 1830s, Faraday was happy to advise the Admiralty on whether oatmeal on prison transports was contaminated, on new methods of treating dry rot, and on the perennial problem of the efficacy of using lightning rods on ships (he also advised the East India Company on protecting their gunpowder stores from lightning). In the 1840s, he advised on possible naval uses of the electric telegraph, but by the end of the decade drew the line at commenting on the quality of disinfecting fluids.

But Faraday's best-known advice to the Admiralty related to the proposed attack on the Russian island naval fortress of Cronstadt in the Baltic, 20 miles west of St Petersburg. In the spring of 1854, Britain and France declared war on Russia to prevent it gaining unduly from the decline of the Ottoman empire. This war, which has become known popularly but unhelpfully as the Crimean War, saw the largest deployment of British forces in Europe between 1815 and 1914. Part of Anglo-French grand strategy was to reduce Russian naval power, hence the siege of Sebastopol on the Crimea peninsular and the plan to attack Cronstadt. The latter was also sufficiently close to the Russian capital to remind the Tsar of the allies' ability, particularly Britain's, to project naval power. The Baltic fleet was notable as the first occasion when the Royal Navy deployed steam-powered warships on active service. The fleet was commanded first by Charles Napier and then Richard Dundas. Thomas Cochrane, 10th Earl of Dundonald, an Admiral of the White, was considered as a possible commander, but was clearly regarded as too old (78) and too eccentric.

Nevertheless, not wishing to be left out of the action, Dundonald (on whom Horatio Hornblower was based) proposed, in the summer of 1854, the use of sulphur-filled fireships to attack Cronstadt. This would involve providing a smoke screen (which seems to be the first proposal of what became a standard naval tactic) through which ships filled with 400 tons of burning sulphur would sail into Cronstadt incapacitating or killing the defenders, thus allowing marines to land and capture the fortress. A secret Admiralty committee was formed to consider this proposal and they sought Faraday's advice. Faraday, a strong supporter of the war, was disappointed with the initial allied failures. In his report, he first referred to his observations of Vesuvius (a rare reference to his Continental travels), then analysed scientifically Dundonald's proposition, but provided the damning conclusion that the proposal was 'correct in theory, but in its results must depend entirely on practical points. These are...untried and unknown.' The committee quoted Faraday's report almost in full, recommending that Dundonald's proposal should not be used. James Graham, First Lord of the Admiralty, accepted the recommendation and Cronstadt was never attacked, but the Baltic fleet was successful in bombarding and capturing other bases, which helped ensure that the Russian Baltic fleet and

30,000 soldiers could not reach the Crimea. Historians now see the Baltic operations as critical in forcing Russia to the peace table and conceding almost everything that the allies wanted. A few months after Faraday's criticisms of Dundonald's proposal, a fire in Newcastle burned 2,000 tons of sulphur, and Faraday wrote to the Admiralty pointing out the comparatively little damage caused. Although Dundonald's scheme was not implemented, the file, nevertheless, remained secret and was not declassified until 1946. The advice that Faraday provided on the proposed attack on Cronstadt was completely characteristic of how he worked in such circumstances. Unlike Davy, whose advice was brief and based on science alone, Faraday's advice was inevitably cautious, given at great length, and, when there were operational implications, would always emphasize, as with Cronstadt, the need for practical knowledge before implementation.

He approached conservation of works of art, on which he was frequently consulted, in the same way. In providing advice on the Elgin Marbles in the British Museum, the Raphael Cartoons at Hampton Court, the conservation of paintings in the National Gallery, the preservation of the stone of the new Houses of Parliament, Faraday always took the most cautious line possible, emphasizing the need for practical experience. This is perhaps best exemplified by his service, in early 1856, on the Royal Commission charged to consider whether the National Gallery should remain in the polluted atmosphere of Trafalgar Square, where it was easily accessible to the public, or move to somewhere like South Kensington which was less polluted, but also less accessible. This classic dilemma of conservation versus accessibility (still very much with us) led Faraday to abstain in the final vote on the grounds that the arguments were too finely balanced.

Another type of advice that Faraday was asked to provide was analytical or inquisitorial, designed to find out why something had happened, especially disasters. The most important was the inquiry into the explosion at Haswell Colliery, County Durham, on the afternoon of Saturday, 28 September 1844, when 95 men and boys died, the three youngest being aged only 10. The inquest was convened by the Durham coroner, Thomas Maynard, at the Railway Tavern in Haswell on 30 September. The explosion had occurred during a period of intense industrial unrest in the Durham coalfield, and the solicitor representing the relatives of those killed personally petitioned the Tory Prime Minister Robert Peel that government representatives be sent to the inquest. Peel agreed and, after some discussion, Faraday (with reluctance) and the geologist Charles Lyell were appointed, together (at Lyell's insistence) with Samuel Stutchbury, a mining engineer from the Duchy of Cornwall. In his letter to Maynard explaining the appointments, the Permanent Secretary of the Home Office, Samuel Phillipps, more than somewhat anticipated the inquest's verdict by referring to the explosion as an 'accident'. Furthermore, he made it clear to Maynard that one of the functions of Faraday's and Lyell's presence was to ensure that the 'verdict would be delivered under the best possible recommendation and with the highest sanction'.

On 8 October, Faraday and Lyell travelled to Haswell and the following day the inquest resumed. Lyell, who had originally trained as a lawyer, later recounted that 'Faraday began, after a few minutes, being seated next the coroner, to cross-examine the witnesses with as much tact, skill, and self-possession as if he had been an old practitioner at the Bar.' On 10 October, Faraday, Lyell, and Stutchbury spent seven or eight hours examining the mine. There they investigated its air flows and identified some laxity in the safety procedures. Thus, much to his consternation, Faraday found that he was sitting on a bag of gunpowder while a naked candle was in use: 'He sprung up on his feet, and, in a most animated and expressive style, expostulated with them for their carelessness.' The following day, after retiring for 10 minutes, the jury returned verdicts of accidental death, which Faraday noted with the comment 'fully agree with them'. After generously contributing to the subscription fund for the widows and orphans, Faraday and Lyell returned to London on 12 October.

They submitted their report nine days later. The government initially reacted favourably and published it as a pamphlet. The report made a number of recommendations and contained the novel observation that coal dust had played a major role in the explosion. The recommendations were mainly concerned with the fire-damp, including that it should be drawn away from mines by improved ventilation. However, the mineowners reacted unfavourably because of the costs involved, which put the government in a difficult position. Instead of accepting or rejecting the report (which it had commissioned), Peel tabled it in the House of Commons on 17 April 1845, the day that the second reading of the highly contentious bill to increase the grant to the Roman Catholic seminary at Maynooth was debated – a good day for a politician to bury bad news. This manoeuvre ensured that no further parliamentary notice was paid to the report's contents. So far as Faraday's role as an expert is concerned, this story illustrates that whilst he was willing to provide advice to the state, he would not, however, tailor the content of his advice for political objectives, though that was what Peel had undoubtedly hoped for.

Trinity House

Of all of Faraday's work for the state and its agencies, the position that took the most time and effort was his role as 'Scientific adviser in experiments on Lights to the Corporation' of Trinity House from 1836 until 1865. An indication of its extent may be gained from considering that after 1836 over 17% of his extant correspondence deals with lighthouse matters.

Trinity House, chartered in 1514 (although its origins went back further), was, theoretically, fully responsible for safe navigation round the shores of England and Wales. In practice, it controlled very little until the Corporation was recast by the reforming Whig government of the early 1830s who gave Trinity House £1,182,546 to purchase all private lighthouses and passed legislation which ensured that future funding would come from harbour dues. The Corporation's administration comprised a Master (a political figurehead), a Deputy Master, a Court of Elder Brethren (most of whom were merchant navy officers), and a Secretary. There were also a number of powerful committees, especially the Wardens, By-Board and Lights. However, it was the Deputy Master and the Secretary who ran the Corporation and executed the instructions of the various committees.

With secure funding and the knowledge that it could gain economy of scale with any innovations it introduced, Trinity House embarked on a major programme of lighthouse development, largely, though not entirely, imitating the French lighthouse service. For example, Trinity House began to change from fixed lights to rotating beams of fixed duration and time intervals. Furthermore, they built new lighthouses where none had previously existed and replaced out-of-date lights such as those at St Catherine's on the Isle of Wight and South Foreland in Kent.

In 1834, John Pelly was appointed Deputy Master, and he decided that to undertake its programme the Corporation needed a scientific adviser. In early 1836, Pelly invited Faraday to accept this role, illustrating again that Faraday at the Royal Institution was seen as Davy's successor in applied science. After some discussion, Faraday was appointed at an annual salary of £200. Twenty years later, he was appointed to a similar position at the Board of Trade to advise on colonial lighthouses, for which he received £100 annually. He also played a significant role in the Royal Commission on Lighthouses in the early 1860s and in helping the Birmingham firm of Chance Brothers develop a capability of manufacturing lighthouse-quality glass.

Much of his work for Trinity House was mundane, such as analysing waters and ensuring that the red and white lead (needed to paint lighthouses) was free from adulteration. But there were also areas where Faraday played a key role in developing novel lighthouse technology. During the 1840s, his efforts for Trinity House were directed towards the problem of lighthouse ventilation. He commenced investigating this in February 1841 following a visit to St Catherine's Lighthouse. Faraday's problem was finding a way of removing from the lighthouse lanthorn the products of combustion from the oil lamps that condensed on the outer glass which thus dimmed the brightness of the light. He developed a chimney which carried away these products without reducing the intensity of the light. In this, the products of combustion went up the inner cylinder of a double glass chimney and by a siphon effect were drawn down the space between the inner and outer cylinders and exhausted outside the lanthorn. First installed in St Catherine's Lighthouse, it proved, judging by the graphic description provided by the keeper two years later, very successful. In 1842, Faraday made over the chimney to his brother, Robert, who had become a gas-lighting contractor. He patented it the following year; Faraday's only patented invention. The chimney was successful, and it was installed in other buildings including the Athenaeum and Buckingham Palace, where, as *The Times* noted, Faraday's lamp illuminated Princess Helena's christening.

Faraday also oversaw the first practical use of electrical light. It seems extraordinary, in retrospect, that the initial utilization of electricity for power (as opposed to communication) should be located in lighthouses, which are generally inaccessible and frequently subject to inclement weather. There were two major schemes proposed, one by a Dr Watson and the other by a Mr Holmes.

Joseph Watson's November 1852 proposal basically involved passing an electric current from a battery across a carbon arc. In such cases, Faraday always insisted that the apparatus should be as complete as possible before examination, and Watson was not ready until July 1854. Tests were then undertaken, and Faraday submitted a 4,200-word report to Trinity House. He began by discussing the positive aspects of the light, noting that it would shine for more than eight hours a day for five days; that the light was brighter than current sources of illumination; and that the carbon rods could be changed in less than a second. The problem, he pointed out, was that to keep the cost of the light to a minimum (always a major concern to Trinity House), the chemical products produced by the battery would have to be collected and sold. In turn, this would mean building a large battery room at some distance from each lighthouse to avoid the fumes produced by nitric acid (this problem had caused one man at the test to spit blood, resulting in Faraday suspending it). Furthermore, accommodation would have to be provided for the three battery men required. Faraday pointed out the problems that would arise by having two sets of employees working at a lighthouse; otherwise Trinity House would need to become a manufacturing and commercial body. He also, more technically, noted that the light flickered too much and that its maintenance required skills and aptitude which lighthouse keepers did not possess. The result of this report was Trinity House decided not even to trial Watson's light.

Three years later, in 1857, Frederick Holmes proposed another method of electrifying lighthouses. Instead of using batteries to power the carbon arc, he proposed using a magneto-electric generator (driven by a steam engine) based on Faraday's 1831 discovery of electro-magnetic induction. Faraday's first report, in April 1857, was fairly positive and asked a series of questions which Holmes answered sufficiently satisfactorily that Faraday recommended that his method should be properly trialled, and Trinity House agreed to contribute towards the costs. During 1858, Holmes installed his light in the South Foreland Lighthouse, and on 8 December electric light, with Faraday present, first shone across the English Channel. Thereafter, Faraday visited South

Science and practice

Foreland regularly and his commitment is shown by a visit on 13 February 1860 when he was snowed up and could not reach the lighthouse. At the second attempt, four days later, 'by climbing over hedges, walls, and fields', he finally visited the lighthouse. Faraday attended the meeting at Trinity House to discuss the results of the trial, and it was agreed that it should be trialled in a revolving light. Electric lights were subsequently installed at Dungeness, Souter Point (County Durham), and perhaps elsewhere, but eventually the technical problems involved proved too great and the electrification programme was abandoned until the 1920s. By then, the incandescent light bulb, invented by Joseph Swan, and the central generation of electricity made electrification a practical proposition.

Trinity House attracted proposals by Watson, Holmes, and others because the Corporation had the material resources to test them. Although Faraday's role was officially to comment on ideas put to Trinity House, it is hard to imagine that Holmes's light would have got as far as it did without Faraday's active backing. Faraday's extensive work for Trinity House was not just about earning extra money or putting science to practical use but had, for him, a deeply moral purpose. His view of moral responsibility that the practitioners of science and technology should have towards the demands of society and the state were neatly encapsulated by him at the start of an 1860 lecture:

> The use of light to guide the mariner as he approaches land, or passes through intricate channels, has, with the advance of society and its ever increasing interests, caused such a necessity for means more and more perfect, as to tax to the utmost the powers of both the philosopher and the practical man, in the development of the principles concerned, and their efficient application.

But, being a Sandemanian, linking all this together was a theological purpose. In the conclusion to his report on Watson's light, Faraday commented:

Much, therefore, as I desire to see the Electric light made available in lighthouses, I cannot recommend its adoption under present circumstances. There is no human arrangement that requires more regularity and certainty of service than a lighthouse. It is trusted by the Mariner as if it were a law of nature; and as the Sun sets so he expects that, with the same certainty, the lights will appear.

Aside from the poetic imagery of this passage, the striking comparison that Faraday made between the laws of nature (which God had written into nature at the Creation) and a human technology, emphasizes the moral seriousness with which Faraday regarded his lighthouse work. At the end of a lecture on the electric telegraph in 1858, he explicitly stated his theological views on the link between scientific knowledge and laws with their practical applications. In his view, science applied practically 'conveys the gifts of God to Man'. This was the justification for all the time and effort he devoted to working for the state and its agencies, when, as he doubtless realized most of the time and explicitly some of the time, he could be working to understand experimentally and theoretically the nature of the universe and the laws that God had written to govern it.

Chapter 4
Electricity

Electro-magnetic induction

Faraday's discoveries of electro-magnetic rotations and the liquefaction of gases, made in the early 1820s, whilst important, did not commence a period of sustained research. The tasks that Davy found for him, and others, such as working with the engineers Marc and Isambard Brunel in the unsuccessful attempt to use gas liquefaction as a source of power, ensured that he could not undertake research systematically during the 1820s. Nevertheless, Faraday had many ideas that he wished to explore and during the decade kept a notebook to record them, in which, at some point, he wrote 'Convert magnetism into Electricity'.

Faraday pursued this goal as time allowed during the 1820s. The July 1825 issue of the *Quarterly Journal of Science* contained a very short article by him entitled 'Electro-Magnetic Current' in which he showed that magnetism did not affect current passing through a wire. Towards the end of the year, he undertook two sets of experiments. In the first, which he called 'Electro Magnetic Induction', he tried, unsuccessfully, to produce electricity in one wire by placing it close to a wire carrying a current. Following D. F. J. Arago's observation that a spinning copper plate would deflect a compass needle, in his second set of experiments, which he called 'Electric Induction', Faraday whirled a copper plate beneath an electroscope but did not detect any electricity. Faraday did not publish these negative results, but they are significant in illustrating his explicit search for the influence of an electric current across space. This is why he used the word 'induction' which in electro-statics denoted the influence of an electric charge.

The glass work intervened and, despite one further attempt in 1828, it was not until 29 August 1831 that he first observed induction. He recorded this set of experiments in his laboratory notebook with the simple heading: 'Expts. On the production of Electricity from Magnetism, etc etc'. The apparatus he used comprised an iron ring wound with two coils of wire on opposite sides (Figure 6). On passing an electric current through one coil, he induced a transient current in the other coil before the galvanometer needle returned to rest, while the wire adopted what Faraday called the electrotonic state. When he broke the circuit, the needle registered a transient current in the opposite direction – this was, in effect, the first electric transformer.

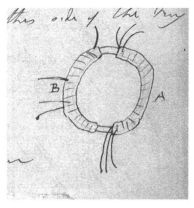

6. Faraday's drawing of the electro-magnetic induction ring

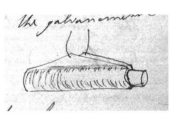

7. Faraday's drawing of the magneto-electric generator

The following day, he repeated and confirmed these results. Three days later, Faraday was in Hastings on presumably a pre-arranged holiday which lasted until the 21 September. Not certain that he had found induction, he told Richard Phillips, the day after his 40th birthday 'I have got hold of a good thing but can't say; it may be a weed instead of a fish that after all my labour I may at last pull up'. Hence he continued experimental work, and on 17 October found how to generate an electric current by moving a permanent magnet in and out of coil (Figure 7).

The crucial day in his experiments was 28 October. Faraday worked with the large array of magnets made by Gowin Knight in the 18th century and loaned to Samuel Christie at Woolwich. He rotated a copper disc between the magnetic poles and generated an electric current (Figure 8). A week later, he was back at Woolwich for what turned out to be the final set of experiments he made before he wrote his paper, the first in his series 'Experimental Researches in Electricity' probably completed by 21 or 22 November. He then went to Brighton, remaining there, 'for health's sake' until 1 December.

Throughout his research, there existed a complex relation between his laboratory notes and published papers. He would draw a vertical pencil line through his manuscript notes to indicate that he had finished with them in the course of writing a paper for publication. But the connection between experiment, notebook, and paper was not straightforward. In the case of induction,

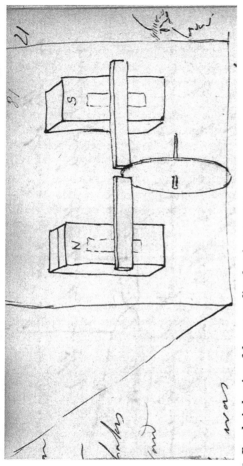

8. Faraday's drawing of the copper disc electric generator

Faraday did not begin with his first experiment, but reordered his account and, furthermore, mentioned Ampère's work 16 times, having not referred to him once in his laboratory notebook.

The paper was read to the Royal Society on 24 November (while he was at Brighton), and 8 and 15 December. Just at this time the Royal Society began introducing a new refereeing procedure which delayed the paper's publication significantly. After the completion of its reading, Faraday wrote to the Parisian savant J. N. P. Hachette describing his results. On Boxing Day, Hachette read this letter to the Académie des Sciences which was quickly reported in various Parisian newspapers. On New Year's Day, *Le Lycée* devoted four columns reporting Faraday's discoveries and claiming that various French savants had made them previously, and on 6 January 1832 the London *Morning Advertiser* published a translation of one of these reports. Faraday, deeply concerned about this, wrote on the 14 January to the Secretary of the Royal Society urging him to try and speed up publication of his paper 'or else these philosophers may get some of my facts in conversation repeat them & publish in their own name before I am out'. This was remarkably prescient of Faraday. Because of these pre-publication accounts, the delay with publication, and the confusion in dates of journal publication occasioned by disruptions due to the 1830 revolution in France, others, such as Leopoldo Nobili in Florence, were given credit for work that Faraday had already done. Indeed, Faraday wrote a pained letter to the *Literary Gazette*, correcting the impressions thus generated, concluding:

> I never took more pains to be quite independent of other persons than in the present investigation; and I have never been more annoyed about any paper than the present by the variety of circumstances which have arisen seeming to imply that I had been anticipated.

On 26 March 1832, Faraday established the mutual orthogonality of electricity, magnetism, and motion, the figure for which

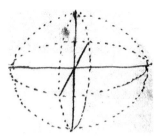

9. Faraday's drawing showing the mutual orthogonality of electricity, magnetism, and motion

illustrates his capability to think three-dimensionally (Figure 9). He had this insight too late for inclusion in series 2 of his 'Experimental Researches', which dealt mostly with terrestrial electricity. Also too late was his novel speculation 'that magnetic action is progressive, and requires time', which he wrote in a sealed note to the Royal Society. The system of sealed notes, established in 1825, allowed Fellows to provide evidence of priority of ideas should the need arise. Faraday was very clear that this was his motive in writing the note, but it seems probable that he forgot about its existence (it was opened in 1937), since during 1832 he switched his attention from electro-magnetism to electro-chemistry.

Electro-chemistry

This switch happened because Faraday wanted to show that electricity generated by magnetism was identical with that derived from other sources – static (known as common), voltaic, animal, lightning, and thermo. As this identity was challenged by his new colleague at the Royal Institution, the Professor of Natural Philosophy, William Ritchie, and also by John Davy, it was imperative for Faraday's future research that it be established firmly. He needed to show that electricity from whatever source produced the same effects (such as electro-chemical, heating,

Table of the experimental Effects common to the Electricities derived from different Sources +.

	Physiological Effects.	Magnetic Deflection.	Magnets made.	Spark.	Heating power.	True chemical action.	Attraction and Repulsion.	Discharge by Hot Air.
1. Voltaic electricity......	×	×	×	×	×	×	×	×
2. Common electricity......	×	×	×	×	×	×	×	×
3. Magneto-Electricity......	×	×	×	×	×	×	×	
4. Thermo-Electricity......	×	×	+	+	+	+		
5. Animal electricity......	×	×	×	+	+	×		

10. Faraday's tabulation of the identity of electricities
+ indicates identities established since first publication

physiological, sparks, and magnetism). He did this both by going through the literature finding those identities already established and by experimentally demonstrating others himself. He collected all this evidence together, tabulated it, showing which identities still needed to be established, and sent it to the Royal Society on 15 December 1832 as series 3 of his 'Experimental Researches' (Figure 10). His continuing concern with this issue is shown by his indicating where some gaps had been filled during the 1830s, by himself and others, when he reprinted the paper in 1839.

In establishing these identities during 1832, Faraday, for the first time, undertook on his own account an extensive series of electro-chemical experiments. Until then, his only work on electro-chemistry had been assisting Davy, when they were on the Continent, with Davy's successful isolation of iodine and his unsuccessful attempts to obtain electro-chemical decomposition from the electricity produced by a torpedo fish. Faraday may have deliberately refrained from experimenting on electro-chemistry out of a desire not to trespass, once again, on what Davy would have regarded as his territory.

By 1832, circumstances had changed, and in the course of establishing the identity of electricities, Faraday had begun to find problems with some of Davy's theoretical views. He found that the electro-chemical action occurred not at the poles (as Davy had theorized), but in the solution itself. As a consequence of this discovery, Faraday began to criticize the two-fluid theory of electricity and instead began to conceive electricity 'as an axis of power having contrary forces, exactly equal in amount, in contrary directions'; hence he thought that electro-chemical decomposition was 'produced by an internal corpuscular action'. In this interpretation, electro-chemical decomposition was caused by the weakening of chemical affinity in one direction which allowed particles in solution to pass out of chemical combination and move to the electric poles (Figure 11).

Furthermore, in a remarkably clever experiment performed on 30 May 1833, he showed that water could be made to behave as an electric pole. All these experiments, which moved electro-chemical theory away from Davy's, forced Faraday, towards the end of 1833, to consider electro-chemical nomenclature; he 'wanted some new names to express my facts in Electrical science'. He discussed this first with the physician Whitlock Nicholl, a Royal Institution Manager. They were both present at the Managers' meeting on 3 December, and two weeks later Faraday in his laboratory

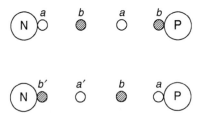

11. **Faraday's idea of electrolytic action**

notebook replaced the word 'pole' with 'electrode', which Nicholl appears to have suggested, as well as the term 'electrolyse' to replace 'electro-chemical decomposition'. These were not enough for Faraday, and on advice he wrote to the polymathic William Whewell about other terms. Whewell, as Faraday knew, had coined terms such as 'Miocene' and 'Pliocene' as names for geological strata for Charles Lyell's *Principles of Geology* (1830–3). After much discussion, by letter, Faraday introduced into science other neologisms including 'cathode', 'anode', 'anion', 'cation', and 'ion'. The last term, which he thought he would not have much use for, arose because Whewell misread Faraday's handwriting after he failed to dot an i.

Wishing to avoid ambiguity in his meaning, precision of language was important for Faraday. He was, however, well aware, as he wrote in paragraph 666 of series 7 of his 'Experimental Researches', 'that names are one thing and science another'. Not only did electro-chemistry force Faraday to rethink the nature of electrical force, he also became strongly opposed to Dalton's theory of chemical atoms, criticizing it at every opportunity. In late June 1834, he breakfasted in Dublin with the mathematician and astronomer William Rowan Hamilton, who was surprised to find that as a practical chemist, Faraday held 'almost as anti-material a view as myself… He finds more and more the conception of matter as an incumbrance and complication in the explanation of phenomena.' What Faraday did not have at this point was any over-arching theory to replace the conventional view, and over the next quarter-century he returned frequently to the problem of finding an alternative.

Nature of electricity

Faraday's understanding that prevailing electrical and matter theories were inadequate, and presumably that some new view was needed, brought about a pause in his publications which lasted just over three years. Series 10 of his 'Experimental

Researches' was dated 11 October 1834, and series 11, 16 November 1837. In January 1834, Faraday indicated in a letter that he already had series 11 and 12 in prospect. What happened to prevent him writing these papers for nearly four years was that during this period he thought about and experimented on the nature of electrical induction and charge and, furthermore, during 1835 spent a considerable amount of time trying, unsuccessfully, to isolate a sample of the highly volatile chemical element fluorine.

His work and thoughts about the nature of charge and of electricity followed two strands. One was the possibility of the existence of absolute charge, the other was the nature of the electric discharge. He meditated on both subjects during four days (3rd, 6th, 9th, and 26th) in November 1835, thinking about a whole range of questions relating to how static and voltaic electricity were linked, how they both related to matter and the nature of induction. For the following two years, he worked alternately on both aspects, but, presumably to avoid confusion, published his results on each topic in separate, but related, papers.

To investigate absolute charge, at the beginning of December 1835, he borrowed a large (31-inch diameter) copper boiler and sought to map the distribution and intensity of electricity in its volume and on its surface. By Christmas Eve, he was confident that his results justified the construction of a 12-foot 'paper box' to map the distribution of electricity on a grand scale. Built in the Royal Institution's lecture theatre during the second week of January 1836, this first "Faraday cage" represented a reversal of Faraday's normal practice of taking experiments from his basement laboratory to demonstrate to the audience in the lecture theatre.

The cube was a 12-foot wooden framed structure covered with wire and paper, mounted on four glass feet, with a small flap about 2 foot square. When charged electro-statically, the cube was insulated from the earth and indeed from the rest of the universe. Faraday 'went into the cube and lived in it' with 'lighted candles, electrometers and all other tests of electrical states'. With this apparatus, he observed that electrical charge was dependent on the electrical state of the observer. This was strong evidence that electricity was a force rather than an imponderable fluid. Furthermore, it disproved, for Faraday, the notion that there existed an absolute electrical charge that could be captured and measured.

Faraday's confidence in the value of his researches received confirmation towards the end of 1836, when the Professor of Mathematics at the university on Corfu (then under British rule), Ottaviano Mossotti, sent him a 34-page pamphlet entitled *Sur les forces qui régissent la constitution intérieure des corps* (1836). Faraday sought Whewell's opinion, writing that Mossotti's 'view jumps in with my notion...that Universal Gravitation is a mere residual phenomenon of Electrical Action & repulsion'. Whewell, in London over Christmas 1836, met Faraday to discuss his work. The following year, in his *History of the Inductive Sciences*, Whewell summarized fairly Mossotti's views, but cautioned as to their 'value and probable success'. Faraday's work was directed towards understanding how all the forces of nature related to each other, not just in understanding electrical action. This he made clear in a lecture on Mossotti's ideas delivered in January 1837; there is, however, no record in his notebook that he pursued such ideas further in the 1830s.

One issue Faraday considered during his meditative days of November 1835 were the observations made by the Professor of Experimental Philosophy at King's College, Charles Wheatstone, when he measured, for the first time, the velocity of electricity. Wheatstone passed an electric current through half a mile of wire with spark gaps at both ends and in the middle. He observed, using the stroboscopic effect of a rotating mirror, that the middle spark occurred later than the sparks at both ends which flashed simultaneously. Faraday thought this a key observation and commented in his notebook that the retardation of the middle spark was probably 'a connecting link between conduction and induction'.

This thought led Faraday to spend much of his experimental time during 1836 and 1837 investigating electrical discharges using variations of Wheatstone's stroboscopic method to observe the development of a spark. He found that when he replaced one of the wires connecting the electrodes in Wheatstone's circuit with water, glass, or some such bad conducting substance, the retardation of the middle spark increased. Faraday conceived of a gradual accumulation of charge in these substances which when it reached a maximum the spark would discharge. He concluded that there existed a unique relationship between matter, light, and electricity. Twenty years later, this provided the theoretical underpinning for spectro-chemical analysis.

In October 1837, he went to Brighton, for nearly four weeks, to write his paper 'in peace & quietness' on the implications of his experiments. Over the previous few years, Faraday had struggled with fundamental physical concepts and he needed to think very carefully about all the issues involved. In series 11, dated 16 November, he introduced two key explanatory concepts. First, he proposed the term 'dielectric' (suggested by Whewell) to denote the electrical state of a non-conducting body between two inductive conductors. Second, he argued that 'common induction was in all cases an *action of contiguous particles*, and that electrical action at a distance...never occurred except through the influence of intervening matter'. He did not use the term 'contiguous' in this sense in his laboratory notebook until January 1838, again illustrating that the relationship between his notebook and published papers was not straightforward. He later expressed some reservations about using 'contiguous' by emphasizing that the particles were merely next to each other rather than touching. Here Faraday tackled the age-old problem of how forces acted at a distance, an issue he would address more fully in the following decade.

Electricity and life

One of the enduring myths in the history of electricity is that Mary Shelley, in her story *Frankenstein, or, the Modern Promethus* (1818), had the very young Victor Frankenstein use electricity to bring his creature, constructed of body parts from a charnel house, to life. Although 20th-century films invariably depicted this use of electricity, she did not, merely referring, vaguely, to 'the instruments of life'. Beyond this technical point, the term 'Frankenstein' came to refer negatively to people who interfered with matters seen as the preserve of providence.

Michael Faraday

That there existed a relationship between electricity and life had been part of electrical studies from the time Luigi Galvani demonstrated that frogs were affected by and produced electric current. Indeed, Faraday had his own frogery in his laboratory where he kept them for use in detecting the presence of minute quantities of electricity, for example in his induction and identity of electricity experiments of 1831 and 1832. Furthermore, when on the Continent with Davy in Montpelier and Rimini, he had experimented on the electric charge produced by torpedo fish.

In late 1836, Andrew Crosse, a gentlemanly electrician of Fyne Court in Somerset, was attempting, using electricity, to form crystals from silica when some living mites of the genus *Acarus* crawled out. This 'Extraordinary Experiment' was first published, under that title, on the last day of 1836 in the *Somerset County Gazette*. Crosse publicized his discovery during early 1837 by, for example, writing to the Bristol Institution. This letter was published in the *Bristol Journal* and from there made its way into the March issue of the national *Gentleman's Magazine*; *Fraser's Magazine* satirized the unnamed Crosse as 'The New Frankenstein'.

One of the other savants to whom Crosse wrote was the comparative anatomist Richard Owen, who received examples of the creatures, and on 17 February 1837 his father-in-law, William Clift, conservator of the Hunterian Museum, exhibited some of these insects in the Royal Institution's library. This was, of course, done with Faraday's permission, but he made it very clear that he gave no opinion one way or the other on the source of these insects. Nevertheless, journalists being journalists, Faraday was reported in a number of papers as having endorsed Crosse's view of their creation. So alarmed was Faraday by these reports that he wrote both to the *Literary Gazette* and *The Times* to disavow them, but in vain. Views incorrectly attributed to him were published in the *Annual Register* for 1837, from which source Harriet Martineau in her *History of England during the Thirty Years' Peace* (1849–50) described Faraday's alleged opinion. She accepted gracefully his correction, but did wonder how one could be sure of historical material if sources such as the *Annual Register* were unreliable.

What is interesting about this episode is not that Faraday sought to set the record straight, but that he did not feel in a position to deny, in public at least, the correctness of Crosse's claim. This contrasts with his private comments, to Christian Schoenbein for example, expressing strong scepticism about Crosse's claims. Such a contrast suggests a strong interest in the relationship between electricity and life, and that he did not want to restrict his options for future research. As with his views on Mossotti's theories, it is clear that Faraday's scientific research was not simply concerned with the technicalities of electricity, but sought to understand and link together all natural phenomena.

Faraday's interest in electricity and life is most apparent in some work that he did on the gymnotus fish the year after the Crosse episode. Like the Mediterranean torpedo fish, the gymnotus, found in South America, also produces a powerful electric shock to kill or stun its prey. In 1834, Faraday asked, unsuccessfully, the Duke of

Sussex to request the Admiralty to acquire a gymnotus from the British colony of Guiana. Furthermore, Alexander Humboldt at Faraday's request provided instructions for its care. In 1835, he applied to the Colonial Office to obtain the fish, but again without success. It was not until September 1838 that a specimen was acquired by the Adelaide Gallery (a venue for popular scientific entertainment), and they permitted Faraday and others to experiment on it for the remainder of the year. Faraday had a saddle made for the fish and with this he mapped the lines of electrical force surrounding it (Figure 12).

Much of Faraday's paper (series 15) on the gymnotus, dated 9 November 1838, related to showing that electric fish produced the same effects as other forms of electricity. From his own experimentation, he could tick more boxes in his table establishing the identity of electricities. But he opened his paper by asserting 'Wonderful as are the laws and phenomena of electricity when made evident to us in inorganic or dead matter, their interest can bear scarcely any comparison with that which attaches to the same force when connected with the nervous system and with life.'

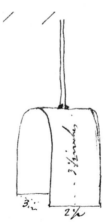

12. **Faraday's drawing of his saddle for the electrical gymnotus fish**

Faraday did not pursue this idea, and indeed put an implicit theological caveat on his investigation: 'We are indeed but upon the threshold of what we may, without presumption, believe man is permitted to know of this matter.' In his Royal Institution lecture on the gymnotus, he went further and 'observed that every remark he might make would be applied only to matter in relation to animal life, not to the principle of life – not to the immaterial spirit'. He was evidently taking care not to have the accusations made about Crosse apply to him.

The interconvertibility of forces

Despite having requested the government in the mid-1830s to obtain a gymnotus for him, Faraday's study of the fish in 1838 was entirely dependent on the chance that the Adelaide Gallery had obtained one for the amusement of their visitors. For the next few years, there would be very little consistency in Faraday's research, both in approach and content. By late 1838, he seems to have recognized that in some sense he had come to a natural conclusion of the research he had commenced in 1831. During late 1838 and into 1839, he collected into a single volume, with a thorough index, his first 14 series of 'Experimental Researches in Electricity', all originally printed in the *Philosophical Transactions*, and published by Taylor and Francis in May.

During the first two-thirds of 1839, there were very few entries in his laboratory notebook. Late that year, he studied the important but difficult subject of the origin of electricity in the battery (which formed the subject of series 16 and 17). In August and September 1840, he worked a little on induction (of which nothing was published), and thereafter there are no entries until June 1842. Then he commenced another opportunistic set of experiments inspired by the observation made by the Newcastle engineer William Armstrong of the generation of frictional electricity by steam jets. This work formed the basis of series 18, read to the Royal Society at the beginning of February 1843, and which he regarded as probably his last paper. The implied belief, that he had reached the end of his research career (in his mid-50s), probably accounted for his decision to publish towards the end of 1844 a second volume of collected papers. This included series 15 to 18 (comprising just under half the volume) and about 25 other papers published in journals other than the *Philosophical Transactions* and thus, as Faraday noted, lacked the harmony of the first volume.

During the three-year period between 1834 and 1837, when Faraday had paused to think about how to proceed with interpreting his results, he had continued experimentation. By contrast, during his second pause in the early 1840s, he undertook very little experimentation and consequently we have little evidence of what he was thinking. However, some of the questions that concerned Faraday were doubtless closely related to those that he considered during his days of meditation of November 1835 and which remained unresolved. For instance, he thought it 'scarcely likely' that iron, nickel, cobalt, and chrome were the only magnetic materials, but that magnetism was a temperature-specific force. Furthermore, although he had argued against the existence of atoms, he still had to provide an alternative theory, which would need to preserve the idea of chemical equivalents. Finally, despite having already shown direct connections between a wide number of forces (electrical, magnetic, dynamical, and so on), he was well aware that other forces needed to be shown to be interconvertible, especially gravity. This theme recurred in his notebook in 1836 and again the following year: 'Compare corpuscular forces in their amount, i.e. the forces of Electricity, Gravity, chemical affinity, cohesion, etc. and give if I can expressions of their equivalents in some shape or other.' Although Faraday did not know it, all these questions, with the exception of gravity, could be investigated with a single research strategy, which will be subject of the next chapter.

Faraday's attempt to relate gravity directly with other forces illustrates his persistence over several decades in holding a theoretical view of the interconvertibility of forces in defiance of experimental evidence. He paved the way for this in the 1830s by showing that electricity was a force (rather than a fluid), so making it like gravitation in that respect. But it was not until 1849 that he specifically turned his theoretical and experimental attention to the question. He had one of his meditative days on 19 March, during which he decided how to approach the problem, his final paragraph commencing: 'ALL THIS IS A DREAM. Still examine it by a few experiments. Nothing is too wonderful to be true.' Faraday started experimentation seriously in August and September when, using their gravitational attraction, he brought together two bodies and attempted to detect if electric currents were produced in them, but found none. Since a further year passed before he sent these negative results to the Royal Society (as series 24), he may have published them reluctantly. He concluded this paper with a statement illuminating his theoretical tenacity: 'The results are negative. They do not shake my strong feeling of the existence of a relation between gravity and electricity.'

Such a strongly stated sentiment meant that it was likely that he would continue to pursue the subject in the future. In a controversial lecture entitled 'Conservation of Force', delivered in February 1857, Faraday argued that the theory of gravitation as generally understood was incompatible with the law of the conservation of force. Interestingly, the young James Clerk Maxwell treated Faraday's views seriously and sympathetically. Such support doubtless encouraged him to continue his experimental work, and in 1859 he undertook another sustained series of experiments, involving dropping a coil of wire from the top of the shot tower in Lambeth, seeking evidence of a direct relationship between gravity and electricity or heat. In May or June 1860, he sent the resulting paper to the Royal Society. George Stokes, one of the Society's Secretaries, seems not to have treated this as a formal submission of a paper, but commented adversely on it to Faraday, who accepted his judgement and withdrew it. Faraday's guiding principle of the unitary nature of all forces could lead him into failure (which he seemed not to mind), but more often led him to success.

Chapter 5
Magnetism, matter, and space

The problem of magnetism

In addition to thinking about why magnetism was specific to so few metals, determining the nature of matter, and expanding his list of convertible forces, other issues intruded in Faraday's life. At the end of November 1839, he suffered some kind of breakdown, the symptoms of which included vertigo, giddiness, and headaches. This illness entailed an immediate visit to Brighton for a month. Throughout 1840, he kept going, though with frequent trips out of town. In December, the Managers said that he had no need to perform his duties until completely recovered, and during 1841 he spent three months in Switzerland (his first visit to the Continent since the journey with Davy), during which he turned 50. Faraday never regained complete health and his letters are full of comments on its adverse state. Nevertheless, like many Victorians, he kept going, and on 15 October 1840 became an Elder of the Sandemanian Church with all the extra duties that position entailed. Finally, during the early 1840s, he was heavily involved with practical work, for Trinity House, the Ordnance Office, and the Home Office. It might be argued that Faraday allowed himself to undertake all these additional tasks because he did not yet know which path of scientific experimentation would lead to the results he desired, nor, indeed, whether such a path existed.

Faraday's resumption of experimentation and the commencement of feeling his way towards resolving his scientific problems began in a lecture he delivered on 20 January 1843, when he posed the following paradox:

> Matter and space query their nature – why space *not a
> conductor* judging by insulators and why it *is a conductor* judging
> by metals – don't know much of matter or space or particles &
> atoms – our province is to determine a power of force & determine
> its laws – that is all.

It was precisely this question which he addressed in another lecture exactly a year later. In 'A speculation touching Electric Conduction and the Nature of Matter', published in the *Philosophical Magazine*, Faraday argued that the question of whether space conducted electricity or not could not be resolved in terms of Dalton's atomic theory. Instead, he proposed that distributed throughout space were points where lines of force met. The physical properties of chemical molecules would be the result of particular combinations of lines of force meeting at a point. In this view, 'matter is not merely mutually penetrable, but each [point] atom extends, so to say, throughout the whole of the solar system, yet always retaining its own centre of force', adding that this 'seems to fall in very harmoniously with Mossotti's mathematical investigations and reference of the phenomena of electricity, cohesion, gravitation, &c'.

Faraday once again sought to understand the universe in its entirety, and his 'Speculation' is perhaps best read as a synthesis of his ideas at that time which would finally overcome the perennial problem of how forces act at a distance. However, he lacked any experimental evidence to justify such a radical theory. The obvious approach to this issue was the problem that he had identified in November 1835 that it would be peculiar if only a very few metals evinced magnetic properties. His 'Speculation' implied that point atoms should be structurally similar to one another. Therefore, if he could show experimentally that magnetism was a universal property of matter, this would supply, for him at least, evidential support for his theory of point atoms; it was to this goal that he now directed his efforts.

Unfortunately, it was exactly at this time that Faraday's work for the state was at its most intense, which he believed impeded his research. As he told the Admiralty Hydrographer, Francis Beaufort (of the wind-force scale), at the beginning of November 1844:

> I have been so long delayed from my own researches by investigations
> & inquiries not my own that I must now resume the former[.] They
> are of great importance & yet a succession of references & subjects
> from government…has kept me from them for the last six months
> and I am now therefore resolved to shut my eyes to all but them.

However, Faraday may not have been as confident as this suggests, since at the end of the following month he told Dumas that he had 'little hope' of producing more scientific knowledge.

Faraday's first attempt at universalizing magnetism was exploring his November 1835 idea that it might be a temperature-specific phenomenon. It was well known that iron lost its magnetic properties as it was heated, so it was a reasonable hypothesis to examine whether metals gained magnetic properties when cooled. Faraday first needed to find out what was the lowest temperature he could achieve artificially in his laboratory. This led him for nine months, starting in May 1844, to pursue his old interest in liquefying gases, using a high-pressure pump. At the beginning of this research, he liquefied olifiant gas (shortly to be renamed ethylene, C_2H_4). This meant that he had reached -154.7°F (-103.7°C), which allowed him to liquefy a few other gases, and he sent these results to the Royal Society in January 1845.

Having achieved such low temperatures, in May Faraday set about trying to obtain magnetic effects from a wide range of materials.

This work included cooling about 40 materials to -166°F (-110°C) to see if they became magnetic, but, apart from cobalt, none did. He published these results in the *Philosophical Magazine* with the thought that the only difference between iron, nickel, and cobalt in their magnetic properties that distinguished them from other metals was temperature. Once again, Faraday put his fundamental beliefs about the nature of the physical world (in this case, the universality of magnetism) before the results of experimentation.

Magnetizing light and matter

Such determination was well rewarded when he attended the British Association's Annual Meeting in Cambridge held, unusually early, from 19 to 25 June 1845, under the presidency of John Herschel. It is unlikely that Faraday attended it entirely, but he was certainly there on 24 June (when he gave a talk at the evening soirée on electricity and magnetism) and the following day (when he took part in a discussion on mining accidents). But more importantly, he met William Thomson, who turned 21 on the 26 June. Thomson was the star of the Cambridge mathematical tripos examination that year, being placed second Wrangler and Smith's prizeman, and consequently had just been elected a fellow of his college, Peterhouse. While an undergraduate, he had dismissed Faraday's approach to science. In March 1843, the 18-year-old Thomson noted 'I have been very much disgusted with his [Faraday's] way of *speaking* of the phenomena, for his theory can be called nothing else.'

Nothing is known of their meeting in Cambridge other than that they discussed the distribution of static electricity. Their meeting led to a close friendship, conducted mostly by correspondence following Thomson's appointment as Professor of Natural Philosophy at Glasgow University in 1846. But Thomson would call on Faraday when he was in London, and at the 1859 meeting of the British Association in Aberdeen, he arranged for the Faradays to stay with his wife's uncle.

In August 1845, Thomson wrote asking Faraday what effect a transparent dielectric would have on polarized light (the query with which Thomson had closed his paper to the British Association). Faraday replied that he had tried the experiment in 1834, but had found no result, and he referred Thomson to series 8 of his 'Experimental Researches', but added that he proposed to reinvestigate the topic. The opportunity to repeat these experiments arose at the end of August when he was asked by Trinity House to test four powerful lighthouse lamps. By 30 August, Faraday was using a lamp to repeat his experiments on passing light through electrolytes, which he continued into September, but again with negative results.

In this set of experiments, Faraday used a piece of heavy optical lead borate glass that he had made in the late 1820s. When next in the laboratory, on 13 September, he decided to examine whether a powerful electro-magnet exerted any effect on polarized light passing through various media (Figure 13).

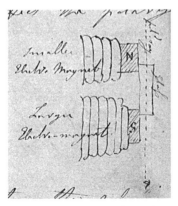

13. Faraday's drawing of the experimental arrangement to show the magneto-optical effect

He observed light transmitted through the heavy glass changed its state of polarization when the electro-magnet was turned on: 'thus magnetic force and light were proved to have relation to each other. This fact will most likely prove exceedingly fertile and of great value in the investigation of both conditions of natural force.' Faraday was fortunate in having access to this piece of glass, since otherwise he would not have discovered the magneto-optical effect. His discovery was thus contingent on his practical work both for the Glass Committee and for Trinity House – thereafter he viewed the glass project more favourably.

The transparent bodies which displayed the magneto-optical effect he called 'dimagnetics', an analogy with dielectrics. Faraday had made two discoveries. First, that somehow light and magnetism were connected and, second, that glass has been affected by magnetism. The latter was of greater relevance to his search for universal magnetism, and he soon turned his attention to showing that the glass was directly susceptible to magnetic force, failing to find any effect on 6 October. Undeterred, he obtained, from the shipping merchant Charles Enderby, half an anchor link. This he made into a giant horseshoe electro-magnet wound with nearly 522 feet of wire and weighing 238 pounds. This became operational on 3 November, and the following day, when he hung a piece of heavy glass between its poles, he observed, when he turned on the magnet, the glass aligned equatorially between the poles:

> I found I *could* affect it by the Magnetic forces and give it position; thus touching dimagnetics by magnetic curves and observing a property quite independent of light, by which also we may probably trace these forces into opaque and other bodies.

Within a week, Faraday had taken full advantage of this discovery and had found more than 50 substances susceptible to magnetic force, exhibiting the behaviour of either iron or heavy glass; gases, however, eluded him. It was in this context that Faraday introduced the term 'magnetic field' into natural philosophy, first in his laboratory notebook on 7 November 1845 and then, with no ceremony, as if it were the most natural term in the world, 11 times in series 20 and 21, his first papers on diamagnetism.

Faraday realized very early on that with magneto-optics and diamagnetism (as he spelled it after correspondence with Whewell), he had made discoveries of fundamental physical significance. In a mid-November letter to Schoenbein, he gave a very brief account of his work, remarking: 'You can hardly imagine how I am struggling to exert my poetical ideas just now for the discovery of analogies – & remote figures respecting the earth Sun & all sorts of things'; in more precise English, he expressed this sentiment at the end of series 21. This, once again, illustrates Faraday's concern with cosmic issues stemming from his painstaking laboratory experimentation.

News of Faraday's discoveries spread quickly and widely, but it was unclear from early newspaper reports of an informal announcement that he made to a General Meeting of the Royal Institution what precisely he had done. For example, Herschel tried to claim some kind of priority, but admitted 'He who proves discovers'. To make clear what he had discovered, Faraday adopted two strategies during 1846. First, he sent out samples of heavy glass to a number of savants throughout Europe so that they could replicate his experiment and, second, he invited a small number of people into his basement laboratory to witness his magneto-optical experiments and later those on diamagnetism.

Faraday had achieved his goal of making magnetism a universal property of matter. He now saw himself in a position to argue strongly for his view of the nature of space, force, and matter. This he did in a lecture delivered in early April 1846 which included his 'Thoughts on Ray-vibrations', published in the *Philosophical Magazine*. Here he dismissed the notion of Daltonian atoms vibrating to produce light waves or radiant heat in an aether (required, by conventional physics, to transmit them). Instead, he argued that distributed throughout space were lines of force and their vibrations produced light. Where lines met, ponderable matter was observed – Faraday was completely clear that one could perceive '*matter* only by its powers'. In sum:

> The view which I am so bold as to put forth considers, therefore, radiation as a high species of vibration in the lines of force which are known to connect particles and also masses of matter together. It endeavours to dismiss the aether, but not the vibrations.

In Faraday's view:

> The smallest atom of matter on the earth acts directly on the smallest atom of matter in the sun.

Magnetizing the atmosphere

Although Faraday had asserted that all matter possessed magnetic properties, he had not during 1845 demonstrated that gases possessed them. Two years later, his attention was redirected towards the issue following the announcement that the Professor of Experimental Physics at Genoa University, Michele Bancalari, had affected the behaviour of flames by magnetism. By October 1847, Faraday had received a description of these experiments by Francesco Zantedeschi. On 23 October, Faraday verified Bancalari's results – and later expressed surprise that he had 'failed to observe the effect years ago'. He then showed that gases, including carbon dioxide, were affected by magnetism. During the next month, Faraday spent two or three days a week working in the laboratory on gases, all of which he found possessed magnetic properties.

These properties, Faraday discovered, depended on the temperature of the gases and the gaseous medium in which the experiments were conducted. Furthermore, he could only observe the relative magnetic and diamagnetic state of gases. Of these findings the most important was his observation that oxygen behaved magnetically, rather than diamagnetically. He published his results, together with a translation of Zantedeschi's report, in the December *Philosophical Magazine*. Towards the end of the paper, Faraday speculated, as was his wont, about the broader implications of his discoveries. He suggested that as the atmosphere's temperature varied with height, it might be affected differentially by the earth's magnetism; however, he did not then pursue the implications of this.

He returned to the subject after receiving a letter at the end of September 1849 from Joseph Plateau (posted the previous March) describing an experimental approach to determine the absolute magnetic properties of gases. This involved observing the behaviour of gases near a magnet viewed through a telescope (which Faraday borrowed from James South) from a distance of 50 feet, but he could not make any useful observations.

Nevertheless, Faraday's interest was aroused, and he returned to the subject at the beginning of 1850. His first move was to see whether the volume of a gas was affected by magnetism. Hence he constructed a strong container, 1/60th of an inch wide and 2.4 inches square, which he placed between the poles of his powerful electro-magnet, and later of the even more powerful electro-magnet owned by the Pharmaceutical Society, but found no effect. He then filled soap bubbles with gases which he found were affected by magnetic fields. As soap bubbles do not last long, he filled some very small glass tubes with gases and confirmed that oxygen possessed strong magnetic properties.

In the summer of 1850, Faraday rented a house in Norwood and commuted between there and the Royal Institution two or three times a week. On 23 July 1850, he recorded in his laboratory notebook the basic idea of what he called 'atmospheric magnetism', a subject he explored for the remainder of the year. Since the magnetic strength of oxygen weakened as its temperature increased, 'Perhaps the cause here of the daily [terrestrial magnetic] variation and even of the larger annual variations' might be found. Faraday suggested that the source of variation of terrestrial magnetism was linked to the warming and cooling of the oxygen in the atmosphere due to night and day and the changing seasons. In Faraday's view, since two-ninths of the atmosphere by weight was made up of highly magnetic oxygen, he found it 'an impossible thing to perceive' that its changes in temperature did not affect its magnetic condition, and thus what was measured as magnetic variation. He mentioned this idea to Whewell, though he asked him to *'keep this to yourself'* and throughout the summer, Faraday developed it, telling Whewell that it 'works out beautifully'.

This idea brought him into direct contact with one of the largest scientific projects of the Victorian British empire, the establishment across the globe from the 1830s onwards of observatories to collect data on magnetic variation. Observatories, under the general direction of Edward Sabine, were established at Greenwich, Toronto, Hobarton (modern Hobart), St Helena, Cape of Good Hope, and Singapore. There was also an international effort, originally proposed by Alexander Humboldt, with data available from observatories at St Petersburg, Washington, and elsewhere. While huge quantities of data had been collected (and published in enormous quarto volumes) since the start of the programme, very little effort had been devoted to developing a theory to account for magnetic variation, and it was to this that Faraday turned his attention.

He experimented on his hypothesis and analysed the data from a number of magnetic observatories, including unpublished data provided by Sabine. Faraday showed, using mainly graphical methods, in two long papers (series 26 and 27) in the *Philosophical Transactions*, that there existed a correlation between the temperature of the atmosphere and magnetic variation. The

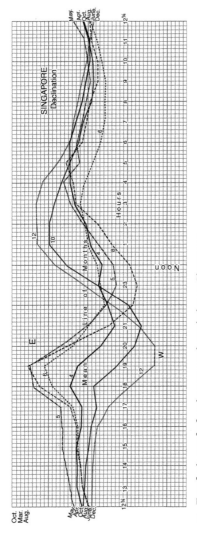

14. Faraday's graph showing magnetic variation at Singapore

reception of this work by some, such as Herschel and Humboldt, was equivocal, but others were more enthusiastic.

Faraday's theory of magnetic variation was completely incorrect (hence the lack of attention it has received), and the concept of atmospheric magnetism did not catch on. Nevertheless, it does illustrate several aspects of his approach to understanding the world. Most interestingly, in this case, is Faraday's unique (for him) use of data reduction, albeit using the visual method of plotting data on graphs. His notebook does not reveal how he did this, but it is clear from the graph (Figure 14) of variation at Singapore (one out of seven that he published) that he must have undertaken a good deal of number-crunching. More usual, by this time, this work shows a concern for taking knowledge produced experimentally in the laboratory and applying it far beyond its walls, in this case to the entire globe.

Mathematizing the field

Faraday's work on atmospheric magnetism formed part of a broader effort to elaborate his field theory and establish the physical reality of lines of force. This continued what he had started in the 1840s on diamagnetism and on which he published new results in 1850 and 1851. Here Faraday adopted the term 'paramagnetic', following correspondence with Whewell, to distinguish materials that behaved differently in a field from diamagnets. To reinforce his idea of the reality of lines of force, Faraday fixed iron filing diagrams in waxed paper, and in late 1851 sent them to his friends. It was this reality he argued for in two papers in 1852 when he presented the iron filing diagrams as evidence for the existence of the field.

Faraday's theory, which he saw as fundamental to understanding space, force, matter, and indeed the universe in general, suffered from one major defect in that it sought to replace well-established, quantifiable mathematical theories with a highly qualitative theory. Hence his views, especially those questioning the existence of the aether, did not go unchallenged, for example by Airy.

The reason why Faraday's theory was adopted in Britain so quickly after he had developed it was that it provided a theory of long-distance telegraph signalling, a crucial issue with the prospect in the 1850s of constructing a telegraph cable from Ireland to Newfoundland across the Atlantic Ocean – the largest, most expensive, engineering project undertaken in the British empire during that decade. The main problem with such a long cable was the phenomenon of telegraphic retardation. That is, in a long cable, a clear signal transmitted was rendered unreadable at the receiving station. First observed by Werner Siemens when he built a cable from Frankfurt to Berlin (some 260 miles) in 1848, Faraday observed it independently in the autumn of 1853 (although he quickly reported Siemens's priority). Then he conducted, with others including Airy, experiments which linked, in series, the telegraph cables between London and Manchester, thus producing a wire 1,500 miles in length. These experiments showed the phenomenon of retardation. To overcome this, essentially what Faraday proposed was to view the cable, insulated with gutta percha, as an electro-static storage jar (admittedly rather long and strangely shaped); one consequence of this proposal was that yet further evidence was provided for the identity of electricities.

This idea was taken up by Thomson who, combining Faraday's theory with a Fourier mathematical analysis (developed for understanding heat transfer), derived equations that described the state of the electric charge at the far end of a cable. At a meeting on the Atlantic cable at the Institution of Civil Engineers in early 1857, Faraday asserted that it should be possible to pass a signal through the cable every two seconds. This was sufficient for the commercial backers of the Atlantic cable and the following year it was laid, although it soon ceased functioning for other reasons. Faraday's approach to science, the field theory, the unity of forces, in this case of two different forms of electrical force, demonstrated that even the most recherché aspects of his work had practical, and in this case relatively immediate, application.

Furthermore, Thomson's success in mathematizing an aspect of Faraday's qualitative theory suggested that his entire theory might be amenable to mathematical analysis. The mathematical natural philosopher who would do this was James Clerk Maxwell. Born the year Faraday discovered induction, Maxwell was second Wrangler in the mathematics tripos examination at Cambridge in January 1854. The following month, he asked Thomson what he should read of Ampère and Faraday on electro-magnetism – a subject not covered by the tripos. Thomson's reply has not been found, but by November they were discussing Faraday's field theory. According to Maxwell's later recollections, he had spent the intervening period reading Faraday's papers rather than mathematics on the subject – that is, he did not then read Ampère's papers which he studied during 1855. He was evidently more impressed with Faraday, telling Thomson in September that he had appropriated Faraday's general notions about lines of force. And it was on this subject that Maxwell read a paper to the Cambridge Philosophical Society in December 1855 and February the following year. It was published in late 1856, and Faraday received his copy early in 1857, which seems to have been their first contact.

Faraday was delighted, and a little alarmed, with Maxwell's approach to his ideas and the way in which it gave his theory the same status as action at a distance theories: 'I was at first almost frightened when I saw such mathematical force made to bear upon subject and then wondered to see that the subject stood it so well.' He returned to this theme later in 1857 when he asked Maxwell why mathematical conclusions may 'not be expressed in common language as fully, clearly, and definitely as in mathematical formulae…translating them out of their hieroglyphics, that we might also work upon them by experiment'. Maxwell's reply, if any, has not been found. But Faraday's delight at Maxwell's achievement was expressed in letters to Sabine and to a niece – a very rare reference to scientific matters in family correspondence: '[Maxwell] wrote a paper on "Faraday's lines of force" and proved their mathematical correctness.'

By this time, Maxwell was Professor of Natural Philosophy at Marischal College, Aberdeen. While there, he did not pursue his electrical studies, but in 1860 he was made redundant following the merger of the two Aberdeen colleges, and in July moved to be Professor of Natural Philosophy at King's College, London. It is clear that Maxwell now came under Faraday's personal as well as intellectual influence, but, because they were in such close geographical proximity, it is impossible to be precise about the nature and extent of their relationship. Maxwell delivered a lecture at the Royal Institution and they would have met at the Royal Society.

Whatever their precise relationship, Maxwell returned to studying electricity and magnetism, the outcome of which was his 1865 paper 'A Dynamical Theory of the Electro-Magnetic Field'. In this, he commented:

> The conception of the propagation of transverse magnetic disturbances to the exclusion of normal ones is distinctly set forth by Professor Faraday in his 'Thoughts on Ray Vibrations'. The electromagnetic theory of light, as proposed by him, is the same in substance as that which I have begun to develope in this paper, except that in 1846 there were no data to calculate the velocity of propagation.

Following his resignation from King's the same year (Maxwell was unfortunate with his university posts), he began work on his *Treatise on Electricity and Magnetism* (1873). This described mathematically the phenomena of electricity and magnetism then known. Maxwell had long viewed Faraday's work as 'the nucleus of everything electric since 1830', and broadly his book can be viewed as a mathematization of Faraday's work. That Maxwell could do this was because he identified Faraday's approach as essentially geometrical – a point he made explicitly in an 1873 profile of Faraday.

However, Maxwell did not follow Faraday in rejecting the aether; indeed, he developed a highly mechanical structure for the aether. Such a typical "British" form of physics found increasing disfavour on the Continent, most trenchantly expressed in Pierre Duhem's *La théorie physique son objet et sa structure* (1906). Maxwell's work, especially his mathematical analysis, helped spread knowledge and acceptance of field theory in Europe. Heinrich Hertz, for example, working in the late 1880s at Karlsruhe University, demonstrated experimentally the existence of electro-magnetic waves entailed in Maxwell's mathematical interpretation of Faraday. Hertz's work formed the scientific basis for the invention of wireless in the following couple of decades by, among others, Marconi, Lodge, and Popov.

In the mathematical hands of Hendrik Lorentz, Professor of Physics at Leiden (1878–1912), and of Albert Einstein, Professor of Physics at Berlin (1914–33), field theory became and remains one of the cornerstones of modern physics. Both Lorentz and Einstein ignored the mechanistic model of the aether proposed by Maxwell and, apparently not knowing of Faraday's rejection of the aether, tended to bracket them together in the British mechanical tradition, whilst realizing the significance and originality of field theory. Einstein, 'who took in Faraday's ideas so to speak with our mother's milk', wrote in 1936 that 'the electric field theory of Faraday and Maxwell represents probably the most profound transformation which has been experienced by the foundations of physics since Newton's time'. Such praise ensured Faraday's and Maxwell's enduring fame in the scientific community.

Chapter 6
Faraday as a celebrity

Lecturing

Faraday's celebrity during his lifetime stemmed from a number of factors, the chief of which, for the general public, were his lectures. When he first joined the Royal Institution in 1813, one of his main tasks was assisting lecturers with their experimental demonstrations in the theatre, and he resumed these duties following his return from the Continent in 1815. The morning lectures for medical students (mostly from the nearby Windmill Street medical school and later from St George's Hospital) were delivered by Brande, while the afternoon lecture courses were for a more general audience. These latter were delivered by a wide range of savants including the engineer John Millington, the poet Thomas Campbell, the architect John Soane, the musician William Crotch, and Peter Roget (of Thesaurus fame). Faraday thus had the opportunity to see at close quarters some of the best lecturers at work, and in a series of letters to Abbott, he reflected on the requirements for high-quality lectures.

Faraday's earliest lectures were delivered to the City Philosophical Society between 1816 and 1818, but not until 1824 did he begin lecturing at the Royal Institution. This opportunity arose because from 1823 onwards Brande became increasingly connected with the Royal Mint and in 1825 was appointed Clerk of the Irons and

Superintendent of Machines at an annual salary of £700, much higher than the Royal Institution could offer. Brande could not therefore maintain the same commitment to the Royal Institution, especially in regards to the morning lectures. Thus, on 7 December 1824, Faraday delivered his first lecture in a course of 19 morning lectures on the metals. In the audience was Roderick Murchison, then changing career from fox hunting to geology. He was clearly not impressed with Faraday's first effort, noting that he mainly read, was too diffident, and used too many colloquialisms.

The reduction of Brande's role within the Royal Institution enabled Faraday to come to the fore. On 7 February 1825, he was appointed Director of the Laboratory; at the end of the year, he was relieved of the task of assisting lecturers. A number of lecture and laboratory assistants were appointed to replace him until Charles Anderson took over in 1832. In December 1826, Faraday was appointed Assistant Superintendent of the House. This position involved overseeing the fabric of the building, from leaking roofs to drains.

It also included supervising the erection of the 14 columned Corinthian façade of the Royal Institution in 1837–8. Based on the Tempio di Antonino in Rome, which Faraday saw when there, the new frontage made a major public statement that something different happened in the Royal Institution compared with the clubs, hotels, and publishers elsewhere in Albemarle Street. So committed was Faraday to this project that he donated significantly to the façade fund.

Faraday's new roles meant that by the mid-1820s, he was the leading employee of the Royal Institution. One of the first things he did was to address its immediate financial problems. In the mid-1820s, two new courses of lectures were established. First, what would become called Friday Evening Discourses, which Faraday helped to found, and the Christmas Lectures for juveniles. This latter series was a development of the afternoon lectures

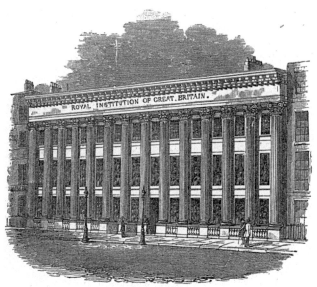

[Royal Institution, Albemarle Street.]

15. Façade of the Royal Institution

that had been delivered since the Royal Institution opened and Faraday's early role in them is unclear, though he would deliver 19 series.

Faraday's Christmas lectures became very popular and publishers, realizing this, sought to persuade him to allow them to be printed, reportedly offering him almost unlimited terms for the copyright. Convinced of the impossibility of turning live lectures, with a large number of experimental demonstrations, into print form, Faraday declined all such invitations until the 1860s. Even as late as 1859, he wrote to one publisher about his rejection of an earlier offer:

> [It was] proposed to take them by short hand & so save me
> trouble – but I knew that would be a thorough failure. Even if
> I cared to give time to the revision of the M.S. still the Lectures

without the experiments & the vivacity of speaking would fall far
behind those in the lecture room as to effect.

Shortly afterwards, he changed his mind and permitted the young
chemist and journalist William Crookes to publish his last two
series in the weekly *Chemical News* and then in book form.
Faraday's final course of Christmas Lectures, *The Chemical History
of a Candle*, delivered in 1860/1 (using notes recycled from 1848/9)
was published in 1861. This is arguably the most popular science
book ever published; never out of print in English, it has been
translated into many languages including French, Japanese,
Polish, and Bulgarian.

Friday Evening Discourses were lectures that only Royal
Institution Members and their guests could attend. Faraday ran
them until 1840 when, following his breakdown in health,
he handed over their administration to John Barlow, shortly to
become Secretary of the Royal Institution. Faraday was assiduous
in ensuring press attention was directed to the Discourses;
reporting, in his view, was clearly different from publishing.
By 1836, he listed the editors of newspapers or journals to whom
invitations were sent, including influential general weekly
periodicals such as the *Athenaeum* and *Literary Gazette*. The
editor of the latter, William Jerdan, Faraday appears to have
known quite well, and certainly by 1828 was providing him with
reports of Discourses. In these and other periodicals, reports,
sometimes quite lengthy, were published of the Discourses. As a
result, Discourses proved an almost instant success, as is evinced
by the annual number of new Members, which grew from an
average of 10.6 in the period 1821 to 1825 to 65 in the following
5 years.

From 1825 until his final Discourse in 1862, Faraday delivered 127
or just under 20% of the total. Whatever problems Faraday had
when he began lecturing disappeared with the Discourses. This is
clear both from attendance figures and by numerous comments

from his audience. The Royal Institution did not start collecting data on audience numbers until 1830. Figure 16 represents the size of Faraday's audience for each Discourse from 1830 until 1862 for which data exists; on three occasions, more than 1,000 people crowded into the theatre to see and hear him. Faraday's absolute numbers not only gradually increased over those three decades, and moreover with the exception of 1835 and 1841 (when he delivered no lectures), the average attendance at his Discourses was always higher, sometimes considerably so, than the average for each year's series.

All this statistical evidence for Faraday's increasing popularity is supported by numerous anecdotes concerning his attractiveness as a lecturer. Richard Owen's wife, Caroline, at the end of the 1830s, described him as 'the *beau idéal* of a popular lecturer', whilst the novelist George Eliot (Marianne Evans) commented that they were 'as fashionable an amusement as the Opera'. Faraday was one of those lecturers, who still exist today, who persuaded their audience

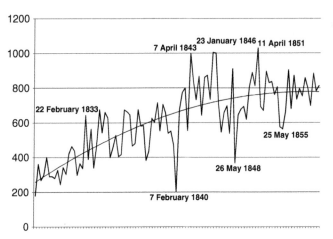

16. Attendance at Faraday's Friday Evening Discourses, 1830–1862

that whilst listening they understood what he said, but who afterwards would have been unable to provide a coherent account of the subject. Nevertheless, the social nature of the Discourses, with Faraday frequently performing and constantly hosting, drove home the consistent message of the value of science to society and culture, and contributed significantly to the general appreciation of the value of scientific knowledge in Victorian Britain.

His success with the Discourses was undoubtedly a major contributory factor in the Royal Institution wishing to elect him President in 1864 in place of the Duke of Northumberland, whose health was failing. The offer provoked a major crisis and despite his wife Sarah's wish that he should accept, Faraday declined, as he saw himself as the Royal Institution's servant, not its master. Indeed, the crisis also led to his relinquishing other tasks, such as the Sandemanian Eldership.

Science and the public

Faraday's lectures, their publication in various forms, and his general celebrity played a vital role in sustaining the Royal Institution's position as the major venue for scientific education, entertainment, and enjoyment. As a result, Faraday became a public figure beyond the narrow confines of the scientific community. He was widely seen as a leading advocate for science and, as such, he attracted both welcomed and unwelcomed attention. He was willing, where necessary, to engage publicly, both pro- and re-actively, with scientific issues of general interest, many of which were controversial in nature. The most controversial were the award of a Civil List pension to him in the mid-1830s and the issue of table-turning in the 1850s and 1860s, both of which were reported in the general press.

At the prompting of James South, Lord Ashley suggested to Robert Peel, Tory Prime Minister from December 1834 until April 1835, that Faraday should be awarded a Civil List pension of £300 annually. Peel was on the point of granting this when he was turned out of office by the Whig Lord Melbourne. Melbourne renewed the offer in October 1835, but at an interview in Downing Street he referred to pensions as 'humbug' with a participle that Faraday's notes describe as 'theological', whereupon Faraday walked out of the meeting. What happened next is not entirely clear, but it appears that the natural daughter of William IV, Mary Fox, who knew South, persuaded her father to put pressure on Melbourne to send Faraday the written apology he had requested. This Melbourne did, but someone leaked the story to the Tory press which, needless to say, had a field day. As a result, the Tory *Fraser's Magazine* published a jokey profile of Faraday which, though it contained many inaccuracies, suggests the author had some knowledge of Faraday's life, including that he was a Sandemanian, 'whatever that may signify'. Both Faraday and Melbourne sought to distance themselves from the controversy, and at one point Faraday contemplated refusing the pension again; the King, however, signed the authorization at the end of 1835.

As Faraday's celebrity grew, he received the usual crank letters that are an occupational hazard for leading members of the scientific community. Such letters were invariably written by very obscure figures proposing impossible flying or perpetual motion machines or chemicals with wonderful new properties. While such letters could be safely ignored, as already being out of touch with the state of scientific knowledge, there were a number of issues where this was not the case and which were keenly contested. One such was the relationship of electricity and life exemplified by Crosse's *acarii* (discussed in Chapter 4) where Faraday had sought to ensure accurate reporting of his views. Although Faraday did not employ the contemporary but little-used term 'pseudo-science', he eventually became highly dismissive of mesmerism (or animal magnetism) and spiritualism, both of which he was forced to pay some attention to because of his public reputation.

Mesmerists claimed that they could change mental states by passing magnets over the human body. This was proposed by Franz Mesmer and taken up by Karl Reichenbach. In 1838, Faraday attended some mesmeric demonstrations by John Elliotson, Professor of Medicine at University College. At this point and into the 1840s, Faraday did not dismiss mesmerism. For instance, in 1844 he might have accepted an invitation from Isambard Brunel to attend a mesmeric session, but had other commitments. However, after his discovery of diamagnetism, which made magnetism a universal property of matter including living matter, he seems to have become dubious. The Assistant Secretary of the Royal Society, Walter White, noted in his diary on 14 March 1846 Faraday's comment that he was:

> not disposed to place faith in the magnetic experiments of Reichenbach, and says that, as of mesmerism, so he cannot believe in them until their law is found to be of invariable application, until they can mesmerise inorganic matter or a baby, who cannot be supposed to be a confederate. He has lost much time in the enquiry without any satisfactory results.

This suggests that Faraday had once taken mesmerism seriously, but no longer. But following his publication of diamagnetism in 1846, he received letters which discussed mesmeric phenomena in great detail. For instance, the Exeter surgeon John Parker explicitly linked Faraday's discovery with mesmerism: 'On perusing the 21st Series of the *Philosophical Transactions* containing your Experimental Researches on Electricity, I am delighted to think that you are approaching to a solution of the true Theory of Mesmerism.' No replies from Faraday to any of these correspondents have been found, and perhaps his views are best summed up by his dismissive endorsement 'Mesmeric stuff' on an 1846 letter from the Irish mesmerist Jane Jennings.

Faraday felt no need to state publicly his views on mesmerism, but such silence was not possible with the arrival of table-turning from the United States in the spring of 1853. Throughout London, tables began to turn, levitate, and occasionally fly in and out of windows. It was inevitable that Faraday, as one of the leading scientific figures of the day, would be asked about these phenomena, and indeed he reported receiving 'numerous' requests for his opinion. A primary reason for his involvement was the widespread belief that the tables moved due to some form of electric or magnetic force. At the end of June, Faraday turned his experimental attention to table-turning. He attended two séances at John Barlow's house and devised an indicator which showed that table-turning was due to 'a *quasi* involuntary muscular action' on the part of those placing their hands on the tables and not some known or unknown force, as the turners claimed.

Unlike with mesmerism, Faraday decided to make his position on table-turning public in an article for the *Athenaeum* and a letter to *The Times* describing his experiments and blaming the widespread belief in table-turning on the educational system: 'I think the system of education that could leave the mental condition of the

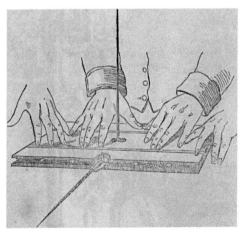

17. Faraday's table-turning indicator

public body in the state in which this subject has found it must have been greatly deficient in some very important principle.' Very quickly, Faraday received letters of support from the scientific and education communities, but he was also severely criticized. The poet Elizabeth Browning privately expressed her outrage accusing Faraday of 'arrogance & insolence'. Faraday was:

> shocked at the flood of impious & irrational matter which has rolled before me in one form or another since I wrote my times letter . . . I cannot help thinking that these delusions of mind & the credulity which makes many think that supernatural works are wrought where all is either fancy or knavery are related to that which is foretold of the latter days & the prevalence of unclean spirits.

By quoting from Revelation 16:13, Faraday indicated that not only had his sense of scientific propriety been transgressed, but so had his Sandemanian beliefs. It was probably this combination that led him to denounce table-turning publicly. He was well aware of the pitfalls involved in entering the fray in such areas, and in the end Faraday came to have contempt for table-turners, especially when they attempted to circulate rumours that he had changed his mind. It is difficult to gauge what effect Faraday's efforts had on the table-turning fashion, other than it was open to wildly differing interpretations. In later years, Faraday's name was invoked both by supporters of research into spiritualist phenomena (such as Crookes) on the grounds that Faraday had taken the subject sufficiently seriously to experiment on it, and by opponents (such as Tyndall) on the grounds that Faraday had provided an entirely satisfactory explanation of such occurrences.

Faraday's criticism of the system of education, which permitted, in his view, table-turning to achieve such widespread popularity, led to the Royal Institution arranging a course of Lectures on Education. He delivered the second lecture, entitled 'Observations on Mental Education', before Prince Albert, on 6 May 1854. Unusually, he wrote the lecture out entirely for reading without any experimental demonstrations since it was to be published by the Royal Institution. He was, according to Sarah, concerned that the lecture would be 'keenly criticized'. In the lecture, much discussed then and since, he returned to the attack on table-turning and gave an autobiographically based account of how the judgement should be educated.

The controversy and legacy of Faraday's interventions on table-turning illustrate that by the 1850s he had become so famous that his views simply could not be ignored. This provided him with the opportunity to set his own agenda which, on occasion, he used. Most famously, on 7 July 1855 he took a steamboat trip on the Thames and was so appalled by the polluted state of the river that he wrote to *The Times* drawing attention to the dangers, especially, he hinted, of cholera. The following week, the satirical magazine *Punch* published a whole-page cartoon depicting Faraday handing his card to Father Thames (Figure 18).

But most of the scientific issues that Faraday tackled in the glare of publicity were in response to requests for advice, and in the 1850s these were mostly on conservation issues, especially the deterioration of the stonework of the new Houses of Parliament constructed between 1837 and 1858. This decay, due to the sulphur content of the London atmosphere, was the cause of such significant political embarrassment that the First Commissioners of the Board of Works (Henry Fitzroy and then, following a change of government, William Cowper) were personally involved. A number of contractors proposed various chemical treatments to consolidate and stabilize the stone. Beginning in about 1856, tests were carried out to determine the effectiveness of these treatments, and in mid-1859, the Board of Works asked Faraday to adjudicate between them. He unambiguously reported that the method invented by Nicholas Szerelmey was by far the best. However, as other contractors were involved, Faraday's opinion was challenged, and he was dragged into an argument that he wanted to avoid, and at one point was threatened with legal action.

FARADAY GIVING HIS CARD TO FATHER THAMES;

And we hope the Dirty Fellow will consult the learned Professor.

18. Faraday handing his card to Father Thames

His response to all this was to reaffirm his belief that Szerelmey's process was the best, but emphasizing, in typical fashion, that the final outcome would only be known after many years of practice. The basic problem was that the stone had been badly quarried in the first place and no amount of remedial conservation would have succeeded in arresting the decay; in the 1930s, much of it was replaced.

For his work for the state, it might be expected that Faraday would, like Lyell, be created a baronet, but the evidence that any offer was made is limited. Faraday always wrote to correct any report that he had been knighted, suggesting he thought the honours system politically corrupt. Nevertheless, this did not prevent him accepting a Commander of the Legion of Honour from Napoleon III (though in what sense the Emperor was not corrupt is a mystery) or a Knighthood from the government of the newly unified Italy (a political cause that Faraday supported), offered via Carlo Matteucci, who served as Minister of Education.

Prince Albert chairing Faraday's lecture on mental education was not an isolated event, but part of a friendship between the two men which was perhaps inevitable given the Prince's interest in science and its applications. Faraday admired the Prince enormously and was deeply affected by his unexpected death at the end of 1861, writing 'We remember him more as a man than a Prince. He exalted his rank far more than it exalted him.' The Prince, elected Vice-Patron in 1843, first visited the Royal Institution in February 1849 when Faraday delivered, in the style of a Discourse, a special lecture on diamagnetism. Thereafter, the Prince frequently chaired lectures, and in 1855 brought his teenage sons the Prince of Wales and Prince Alfred to attend Faraday's Christmas lectures. On 18 March 1858, Faraday was present when the new Chelsea Bridge was opened by the Prince and the Prince of Wales and accompanied them across the bridge. He was probably on the Prince's mind when, just a month later, a grace and favour house on Hampton Court Green became available and Albert suggested to the Queen that Faraday might have it.

For some years, Faraday, to avoid the bustle of London, had frequently rented a house outside the city and commuted to the

Royal Institution. It was thus very convenient for his needs that he should be offered, on a lifetime basis, a house at Hampton Court, reasonably accessible from London. In late April 1858, Faraday, Sarah Faraday, and a niece visited the Queen Anne building located on the River Thames near the Palace. It clearly needed considerable repair and decoration and Faraday was concerned about who would pay for this work, but Buckingham Palace put his mind at rest. By early July, the repairs, estimated at £100, were well under way but, as usual in such cases, more problems were revealed than anticipated, and the Faradays took possession only during the first half of September. Thereafter, Faraday spent increasing amounts of time each year at Hampton Court and seems to have lived there entirely from early 1866, dying there on 25 August the following year.

Images of Faraday

Following a Discourse on 25 January 1839, Faraday announced the invention, by William Henry Fox Talbot, of the technology that would soon be called photography. Faraday's involvement arose because Talbot needed a quick announcement to ensure priority over the very different imaging process invented by Louis Daguerre in France. Faraday was clearly fascinated by Talbot's invention, declaring: 'No human hand has hitherto traced such lines as these drawings displayed.' This fascination continued for the rest of his life. In 1860, he was photographed by Charles Dodgson (better known as Lewis Carroll) and sat for one of the earliest flash photographs (taken during a Discourse in 1864). Indeed, Faraday became one of the most photographed men of the time, although he claimed he could never recognize his image and wondered 'whether all persons are in this curious condition'.

Faraday's interest in photography was part of a broader one in portraiture. He was depicted by many of the leading portraitists of the day, including Henry Pickersgill (1829), William Brockedon (1831), Charles Turner (1838), Thomas Phillips (1841–2),

Alexander Blaikley (1845), Thomas Maguire (1851), George Richmond (1852), and J. Z. Bell (1852); Blaikley and Bell were Sandemanians. Faraday was also sculpted by Edward Baily (1830) and Matthew Noble (1854, Figure 2). Original images, seen only where they were displayed, restricted the numbers who could see them. However, many of Faraday's images were quickly engraved and published as prints for a flourishing market in portraits of celebrities. Faraday was not only involved on the production side of his images, but, as a collector of images, was also a consumer of this visual culture. He kept his portrait collection in two large albums and wherever possible mounted them opposite letters written to him from those portrayed. Faraday had personal contact with most of those who appeared in his albums which thus provide a record of his social world.

Although prints conveyed Faraday's image to a large audience, their publication in the periodical press brought him before an even wider public such as the *Punch* image (Figure 18). Accompanying his profile published in *Fraser's Magazine* at the time of the pension crisis, was a portrait of him by Daniel Maclise showing him standing behind the lecture bench and demonstration apparatus (Figure 19).

He also appeared twice in the *Illustrated London News* when he delivered his 1846 Discourse on the magneto-optical effect and his 1855 Christmas Lecture in the presence of Prince Albert by Alexander Blaikley (Figure 20). Common to all these images is that they depict him carrying out a public scientific role and, furthermore, each of these images was related to a significant news item.

Although the general periodical press reported Faraday's discoveries as he announced them to the Royal Institution and the Royal Society, they were not interested in depicting his place of work. But Faraday was, and during the early 1850s Harriet Moore, a wealthy Royal Institution Member and a friend of the Faradays,

Yours Truly

M Faraday

AUTHOR OF 'CHEMICAL MANIPULATION'.

Published by James Fraser, 215 Regent Street, London

19. Faraday, by Daniel Maclise

20. Faraday delivering the Christmas Lecture on 27 December 1855, by Alexander Blaikley

painted a set of watercolours showing his laboratories and living quarters. These are among the earliest artistic representations of a laboratory and, since they can only have been executed with Faraday's permission, demonstrate his abiding interest in art.

Where this interest came from and how he originally cultivated it is not entirely clear, but from the 1820s he had strong connections with the art world. His brother-in-law George Barnard, a watercolour painter who taught art at Rugby school, studied with J. D. Harding, with whom Faraday was friendly. It was probably through him that Faraday became connected with Joseph Hullmandel, a highly important innovator of lithography. Hullmandel's obituary described him as Faraday's 'pupil' with whom he studied chemistry to improve lithography; Faraday would later deliver three Discourses on Hullmandel's work. Through Barnard, Harding, and Hullmandel, Faraday knew Richard Westall, Clarkson Stanfield, John Landseer, and, by other routes, John Constable, John Martin, and Benjamin Haydon, among other Royal Academicians. Indeed, the only annual dinner which Faraday attended regularly (aside from Trinity House), against his usual rule, was the Royal Academy's.

There is, of course, another meaning to the term 'image' and that is the development and propagation of biographical knowledge of an individual. In Faraday's case, this was closely linked to his visual image. For instance, in 1834 William Jerdan wrote the first published profile of Faraday in his *National Portrait Gallery of Illustrious and Eminent Personages* to accompany a new impression of Pickersgill's portrait. Since Jerdan knew Faraday well in the context of the *Literary Gazette*, Faraday probably provided him with the information and so the biographical details are accurate, though incomplete. Twenty years later, Edward Walford wrote another accurate account of Faraday's life to accompany Maull and Polyblank's wonderfully atmospheric photograph of Faraday holding a magnet. Here there is no evidence of contact between author and subject, but by then the basic biographical facts about Faraday's life were already in the public domain through earlier profiles and biographical dictionary entries.

Faraday was thus a very well-known public figure by August 1867, and following his death there was a flood of laudatory obituaries in all the major newspapers (many reprinted in the regional press) as well as in scientific and religious papers, magazines, and journals. Some of these were not entirely accurate; *The Times*, for example, gave his year of birth as 1794. A public memorial was arranged which took the form of a more than life-size sculpture by John Foley and Thomas Brock depicting Faraday in academic dress holding his induction ring, which suggests that this was already seen as his key scientific achievement. In 1876, the statue was unveiled by the Prince of Wales in the Royal Institution's Grand Entrance, where it remains.

21. Faraday holding a magnet, attracting the world

The passing of such a celebrity prompted three of his friends to write substantial biographies. The first, by Tyndall, was *Faraday as Discoverer* (1868; fifth edition, 1894) based on lectures he delivered at the Royal Institution. He portrayed Faraday as a romantic philosopher, a prophet, and the greatest experimentalist ever. Although Tyndall and Faraday got on well at a personal level, their ideological positions were somewhat at variance. Hence Tyndall made no reference to Faraday's Sandemanianism and was critical of his theoretical work, referring to it as 'dark sayings' as one might expect from someone who, six years later in his Presidential Address to the British Association's Belfast meeting, would declare for a highly materialist interpretation of the world in terms of atoms. In 1870, Henry Bence Jones published his two-volume *Life and Letters of Faraday* which went into a second edition the same year. This portrayed Faraday as being driven by innate genius and differs in a number of important respects from Tyndall. While Tyndall covered the chronology of Faraday's life in a fairly even manner, the first volume of Bence Jones went up to 1830, whilst the second volume covered the rest of his life. Unlike Tyndall, who worked almost entirely from published sources, Bence Jones worked from manuscripts, and the overwhelming proportion of these volumes contain long extracts from Faraday's Continental diary, his letters (including those to Abbott), his experimental notebook, and other unpublished documents. The third biography (published in 1872 with two later editions) was by the chemist John Gladstone, whom he had first met at the Royal Commission on Lighthouses. Like Faraday, Gladstone was a deeply religious man and his book was not so much a conventional biography, but a portrait of Faraday, concentrating to a large extent on his character and Sandemanian religious beliefs. The last major 19th-century biography of Faraday, written by the Quaker physicist and electrical engineer S. P. Thompson, was published in 1898 (second edition, 1901). Thompson was in a position to appreciate, as his predecessors had not, the long-term importance of Faraday's work on electro-magnetism, particularly his enunciation of field theory.

All these books went into two or more editions, suggesting the existence in the latter third of the 19th century of a large readership strongly interested in Faraday. While all four texts acknowledged that Faraday undertook applied scientific work, esecially for the state and its agencies, they did not emphasize its significance. The image that came across from these early studies portrayed Faraday as a lone man of science working with but a single assistant in his basement laboratory making scientific discoveries. It was this image, bequeathed to the 20th century, that would become such a potent emblem of the power of science.

Chapter 7
Faraday in the 20th century

The electrical industry

The centenary of Faraday's birth was celebrated in 1891 with a commemorative meeting held in the Royal Institution chaired by the Prince of Wales. He recollected attending Faraday's Christmas Lectures, more than 30 years before, and the 'admirable and lucid way in which he delivered' them. Lord Rayleigh then lectured on Faraday's contributions to physics, and the following week James Dewar lectured on his chemistry. This was the first of five commemorations held thus far to mark various Faraday anniversaries, namely the centenaries of his discoveries of benzene (1925) and of electro-magnetic induction (1931), the centenary of his death (1967), and the bicentenary of his birth (1991).

The reason why Faraday has attracted more celebrations than almost any other scientific figure is that his image thus projected meshed well with the aims, objectives, and political agendas of various interest groups who organized and funded them, and in this the electrical industry was overwhelmingly pre-eminent. Largely motivated by their desire to provide a scientific pedigree for this new branch of engineering, the interest of the electrical industry commenced in the late 19th century. The year after Faraday's death, the unit of capacitance was named the 'farad', and six years later Charles Siemens named (with Sarah's permission) the first purpose-built telegraph-cable-laying ship the *Faraday*, which would lay about 50,000 miles of cable. These two instances of using Faraday's name illustrate that electrical technology was initially associated primarily with telegraphy, since by naming its main unit and its first ship after him the industry was paying the highest possible compliment to his memory and, at the same time, associating itself with one of the century's most celebrated natural philosophers.

The original professional organization for electrical engineering was the Society of Telegraph Engineers, founded in 1871. The widening scope for using electricity, especially for lighting, led to its changing its name in 1888 to the Institution of Electrical Engineers. In 1899, it began using an image of Faraday on its seal (which its successor body continues to do) and marked the 50th anniversary of its foundation in 1921 by establishing, as its highest award, the Faraday medal. Eminent engineers and scientists such as J. J. Thomson, Sebastian de Ferranti, Lord Rutherford, John Cockcroft, Martin Ryle, and Maurice Wilkes have received this medal.

The expansion of electrical engineering into the vast industry which it sustained required the training, in large numbers, of qualified engineers. S. P. Thompson, as Professor of Electrical Engineering at Finsbury Technical College from 1885 to 1915, taught thousands of students their subject. But this was not sufficient, and in 1890 the Electrical Standardising, Testing and Training Institution was established. The following year, the centenary of his birth, it was renamed Faraday House, and until its closure in 1967 was devoted solely to the purpose of training electrical engineers.

As electrical engineering increased in scope and size in the late 19th century and the opening decades of the 20th, it ran into commercial opposition from already established technologies and industries, for instance steam locomotion in transport and gas for lighting, heating, cooking, and so on. The electrical industry's fundamental commercial problem was the enormous capital investment required to build generating plants, and distribution networks meant that the price of electrical energy was considerably higher than either gas or coal. Different countries tackled the problem in radically different ways. In Russia, for instance, it was the military who from the 1870s took the lead in electrification, with vast technological transfer from the German electrotechnical industry. This use of Russian state resources continued after the Bolshevik revolution and accounts for Lenin's slogan that 'Communism is Soviet power plus the electrification of the whole country'.

In Britain, however, such central state support was not forthcoming, at least initially. In the 1880s, individual generating stations serving small areas, for example Kensington and Deptford, were established by R. E. B. Crompton and Sebastian de Ferranti, respectively. But there were no standards even at the most basic level – Kensington produced DC while Deptford generated AC. Thereafter, electricity production in many areas became the responsibility of the municipal authorities which produced a whole clutch of other problems – an electrical device sold in one town might not work in the next town.

In 1926, the state intervened to sort out these problems. Before the Great War, the government, with the notable exception of telegraphy, was reluctant to interfere directly, beyond regulations relating to safety and environmental controls, in how industries conducted themselves, preferring the action of the market place. The Great War had shown that the state could command industry for a common purpose, although these controls were rapidly dismantled following the armistice in 1918. By the mid-1920s, it was clear that the infrastructural weakness of the British electrical industry meant that the country was falling behind others. Although the electrical industry was not nationalized in 1926, Parliament enacted legislation to create the Central

Electricity Board. The Board's primary responsibility was to establish, to uniform standards, a national grid for the distribution of electricity and a system of large electrical-generating power stations that would replace local generation and distribution facilities. Eventually, this programme was implemented, but electricity was still expensive, and it required sustained marketing, on a limited budget, to promote and extend its use. Various organizations were established to do this including, in addition to the Institution of Electrical Engineers, the Electrical Association for Women and the Electrical Development Association. In 1931, the centenary of Faraday's discovery of electro-magnetic induction provided an ideal opportunity market electricity.

Celebrating induction

It is rare for a specific scientific discovery to be celebrated. The usual occasions for commemoration are the births and deaths of scientific figures, or the publication of a book or a paper. However, the centenary of two of Faraday's discoveries have been celebrated. That of benzene in 1925 was generally restricted to the scientific community and specifically to the Royal Institution. It is perhaps best interpreted as a section of the chemical community attempting to reclaim, from physicists and electrical engineers, Faraday as a chemist. Indeed, the gas industry sought to appropriate Faraday at this time by issuing a leaflet describing him as 'one of the earliest Gas Consultants'.

The 1931 events to celebrate the centenary of Faraday's discovery of electro-magnetic induction were on an altogether different scale. The initial idea emerged from a meeting held towards the end of 1928 between the Director of the Electrical Development Association, Walter Vignoles, and Faraday's successor as Fullerian Professor at the Royal Institution, William Bragg. The result of their meeting was that representatives of all the interested parties were invited to the Royal Institution in early 1929 to discuss what might be done. This meeting was attended by a formidable array of institutions and individuals, including John Reith (Director General of the British Broadcasting Corporation), Lord Rutherford (President of the Royal Society), and Kenelm Edgcumbe (President of the Institution of Electrical Engineers). As a result, two organizing committees were formed, one "scientific" run by the Royal Institution and the other "industrial" organized by the Institution of Electrical Engineers. Although there was considerable overlap and coordination between these committees, they operated independently, were funded separately, and the events they arranged were very different.

Vignoles quickly took control of the Institution of Electrical Engineers committee, which eventually resulted in him resigning from the Electrical Development Association. He organized the exhibition on Faraday and electrification held at the Albert Hall in late September and early October 1931. This event would cost the large sum of £10,000, and after some discussion the Institution of Electrical Engineers agreed to meet this, their Council minutes noting as a result that 'the question of disposing of the Institution's annual surplus would not arise for a year or two'.

By March 1931, the structure of the exhibition had been decided. At the centre of the Albert Hall stood a copy by William Fagan of Foley and Brock's statue of Faraday. The first circle of display cases surrounding the statue were arranged by the Royal Institution and mostly contained Faraday's original apparatus and manuscripts. The outer displays dealt with the modern technologies which were seen to have stemmed from Faraday's discoveries. Most of these cases contained modern devices, although there were occasional original objects and reproductions of historical items. In most instances, the modern exhibits were grouped with the most relevant part of Faraday's work.

A key message of the exhibition was associating electrical technology with modernity. The publicity poster and leaflet were

22. **Leaflet for the 1931 Albert Hall Faraday exhibition, by Edward McKnight Kauffer**

designed by the well-known American-born, French-trained influential avant-garde poster designer Edward Kauffer (Figure 22), while the exhibition's lighting design was expressed in the language of modernity. This emphasis on associating electricity with modernity (an integral part of promoting the use of electricity during this period) in an exhibition celebrating an historic event gave it a Janus-like feeling. Another unusual aspect of the exhibition was that there was no commercial element to it. All the contributors had to remove their company names from the displays. Previously unheard of in electrical exhibitions, some members of the Electrical Association for Women approved of the policy. The exhibition, opened by the President of the British Association (also celebrating its centenary), the imperial statesman Jan Smuts, was judged a success, attracting 50,000 visitors during its 8 days, and producing newspaper reportage valued at around £43,000.

The 'great Commemorative Meeting' organized by the Royal Institution was fixed for Monday, 21 September 1931, and held in the Queen's Hall. Until destroyed by bombing in 1941, Henry Wood's Promenade Concerts (the Proms) had been performed there, and from 1927 broadcast by the BBC. Reith attached so much importance to the Faraday celebrations that he instructed Wood that there would be no Prom that evening, although his orchestra was expected to provide the music for the meeting. At first, the Royal Institution hoped that the Prince of Wales (briefly Edward VIII and later Duke of Windsor) would take part, but they had to settle for Ramsay MacDonald, the first Scottish Labour Prime Minister. MacDonald had strong scientific interests, having intended to be a geologist, and had married the daughter of Faraday's biographer and successor but one to his Fullerian Professorship, John Gladstone.

Unfortunately for everyone involved during the weekend of 18 to 20 September Britain was forced off the Gold Standard – the worst financial crisis of the century. Despite having to see the necessary enabling legislation through both Houses of Parliament, MacDonald honoured his appointment to speak on Faraday, giving as his reason that he wanted to keep the nation calm. The event was broadcast, although the schedule had to be rearranged at very short notice to accommodate an address to the nation by the Chancellor of the Exchequer. MacDonald commenced by emphasizing the importance of Faraday's work: 'Without Michael Faraday there would have been no broadcast.' However, he turned quickly to Faraday's character, stressing his enthusiasm, his simplicity, his lack of ambition, his conscience, and the contradictions in his personality – 'What a miserable personality that has no contradictions!' – suggesting that Faraday's was like a Gothic façade. It is perhaps little wonder that after the strains of the day, MacDonald suffered a nervous breakdown and was taken to Sandwich to recover.

The legacy of 1931

It is impossible to be precise about the effects of the 1931 Faraday celebrations on the development of electrification. They doubtless left a lasting impression on those who visited the Albert Hall exhibition, or who attended the commemorative meeting at the Queen's Hall, or the many other events (including the general release of a film of Bragg telling the story of Faraday's life) that were staged throughout the country and overseas. No recording of the Queen's Hall speeches was kept on the cumbersome shellac records then used (although the Chancellor's crisis speech does exist). Newspaper reports by their nature are ephemeral, but the 22-page special supplement devoted entirely to Faraday issued by *The Times* can be found in the archives of some of those who attended the event, suggesting a long-term interest, and the following year it was translated into French as a 244-page book.

It was through the large number of publications associated with the 1931 celebrations that the impact of Faraday as an emblem of electrification continued to be felt. These publications came in two categories, the popular and the scholarly. Of the latter, Robert Hadfield published a detailed study of Faraday's metallurgical work, but more important was a seven-volume edition of Faraday's laboratory notebooks edited by the General Secretary of the

Royal Institution, Thomas Martin. Published between 1932 and 1936, this was called *Faraday's Diary*. The term 'diary' first seems to have been used by William Bragg in a 1928 lecture, although Martin realized it was no such thing – it is certainly no Pepys or Lees-Milne.

The estimate to publish the notebooks was £2,200, and the Royal Institution arranged a publishing contract with Bell and Son who secured 242 subscribers for the set at £12.12.0 each. Nevertheless, the Royal Institution had to provide a significant subsidy which by early 1934 had risen to £1114.15.2. Under Martin's direction, Royal Institution staff transcribed, proofed, and, where necessary, checked Faraday's descriptions against experiment. On the publication of the first two volumes in November 1932, the Treasurer of the Royal Institution and Government Chemist, Robert Robertson, 'expressed the view that these volumes contained much that was significant to the present day student'. As with the Albert Hall exhibition, there was, so far as Faraday was concerned, this Janus mixture of the past and the present. Since its publication, *Faraday's Diary* has proved an invaluable resource not only for historians of science, but also for scholars in other disciplines, for example in computer modelling of scientific processes and in cognitive psychology; it was reprinted in 2008.

The vast majority of publications associated with the 1931 celebrations were, however, hagiographical and many were ephemeral. The Electrical Development Association published a 12-page pamphlet, *Faraday: The Story of an Errand-Boy Who Changed the World*, and distributed 95,000 copies. As with the rest of the celebrations, this pamphlet drove home the message of the centrality of Faraday's scientific research to modern electrical engineering. Other stand-alone publications issued in 1931 included Rollo Appleyard, *A Tribute to Michael Faraday*; E. W. Ashcroft, *Faraday*; W. Cramp, *Michael Faraday and Some of His Contemporaries*; and William Bragg, *Michael Faraday*, the text of his BBC National Lecture (the precursor to the Reith

Lectures). The industry wanted a "pure" scientific figure for its founder and those writing the biographical accounts obliged. The content of all these texts was derived entirely from the 19th-century biographies of him. Taken together, they portrayed Faraday as a figure who rose from obscurity to fame by his own unaided efforts working by himself (or with one assistant) in the Royal Institution's basement laboratory, laying the foundations of the modern world. Although he was hailed as the founder of electrical engineering, ironically his own practical contributions, for example in lighthouses and telegraphy, were either downplayed or just ignored.

With the reputation that Faraday enjoyed during the 1930s it is not surprising that his life became the subject of a 1942 wartime British Council propaganda docudrama extolling the individualistic virtues of British science against the centralized science practised by the Fascists. The film, staring Walter Hudd and Nova Pilbeam, thus emphasized 'the importance of the system of government under which genius can develop'. Nor should there be little wonder that an impressionable lower-middle-class chemistry undergraduate at Oxford University become interested in Faraday. Margaret Roberts, who studied under Dorothy Hodgkin at Somerville College between 1943 and 1947, was so influenced by the story, as told, of Faraday's rise by his own efforts from obscurity to fame, that it seems to have contributed to the formation of her political views as Margaret Thatcher. As with MacDonald, Thatcher, in a number of speeches, deployed Faraday as a rhetorical image to further her political ends. But for rhetoric to be most successful, it is essential that there be some substance. In the case of Thatcher, in 1982 she borrowed Noble's bust of Faraday from the Royal Institution and, until it was returned during John Major's term of office in 1996, it was the first object visitors saw on entering 10 Downing Street. Thatcher's approach was very similar to the strategy employed by the electrical industry of taking only what was needed from Faraday's image to promote their own interests and emphasize the respectable scientific pedigree of their agendas.

Faraday's recent reputation

That such interest groups could use Faraday in this way during the 1930s with little objection was because no critical history of the science community existed to counter such claims. In the 1930s, all those interested in the history of science were scientists of one form or another who, apart from a few Marxists and American sociologists, could not or would not see beyond the content of science into the wider social and cultural world in which scientific figures worked. It was not until the 1950s that a critical history of science began to develop, which emphasized the need to understand the contexts as well as the content of science. Thus far, historians have not dislodged the simplistic narratives of scientific discovery and progress that are so beloved by scientists, engineers, politicians, and indeed the general public.

Faraday's fortunes at the hands of recent historians and biographers have been mixed. Because of his colossal reputation, he became, along with figures such as Kepler, Galileo, Newton, Huygens, Euler, Lavoisier, Darwin, Maxwell, and Einstein, subject to intense scholarly enquiry as the history of science grew after 1945. The historical quality of such research generally became more sophisticated as history of science moved away from studying just scientific content, often set in a pre-determined philosophical context, to a proper branch of history.

These changes took place over several decades, and they can be traced in the development of Faraday studies during this period. In 1965, L. P. Williams, who moved from chemical engineering to history of science at Cornell University, published a massive biography of Faraday. His main arguments were that from an early stage Faraday was influenced by Kantian *Naturphilosophie* and that his experimental work was guided by a matter theory proposed by the 18th-century Dalmatian Jesuit, Rudjer Bošković, and that Faraday was secretly committed to this theory. Thereafter, quite an industry arose devoted to refuting these highly contentious, undocumented, claims and seeking other 18th-century sources for Faraday's creativity.

There were two fundamental problems with this "intellectual history" approach towards studying Faraday. First, university-educated historians of science at that time found it hard to accept that someone from the artisan class with little formal education could have made such creative contributions to understanding the natural world. He was thus viewed as a "black box" through which 18th-century ideas, such as those of Bošković, Priestley, or James Hutton, were transmitted with little real theoretical input from him; he was merely an experimentalist transmitting the ideas of others. The second fundamental problem was that until the 1970s, experiment was assumed to be entirely subservient to theory. A common view was that experiments were designed within a theoretical framework or, if one followed Karl Popper, used to falsify theory. Either way, experiment by itself was dismissed as "trial and error".

This view of experiment started to change radically in the 1970s. David Gooding, at Bath University, showed, by detailed study of Faraday, that experiment, though related in various ways to theory, could not be reduced to such simplistic purposes, but was an active exploratory process, producing knowledge in its own right. His work, and that of others at the same time, opened up a new range of possibilities for historical investigation beyond Faraday, including how natural phenomena were explored, their visualization, and the role of human agency. Furthermore, this approach allowed links to be made with other already existing areas of the history of science such as institutions and laboratories and through them to wider social and cultural areas such as politics, art, and religion. So far as the latter was concerned, in 1991 Geoffrey Cantor, at Leeds University, published a study of

Faraday's Sandemanianism which located Faraday in one of his key social settings and illustrated how this, together with his belief system, was crucial in his approach to science, in terms of research, communication, and application. Cantor's study was a major blast against some of the scientific community's more vocal elements who denied that religion played any role in modern science. In terms of his place in 19th-century society, Faraday's *Correspondence* (volume one, 1991) edited by Frank James at the Royal Institution, showed his importance to the state, its agencies, and the military, as well as, of course, the scientific and engineering communities.

Such studies should have dismissed once and for all the image of Faraday as a lone genius working away in his basement laboratory making discoveries that were applied by others. However, simplistic ideas which have existed for many decades are notoriously hard to shift. Faraday, in popular terms, continued to be subjected to the same tired old descriptions used in the 1930s. Phrases such as 'the father of electricity' were used by those who should have known better, and indeed this appeared on the poster advertising the exhibition held at the Science Museum in 1991 to mark the bicentenary of Faraday's birth.

In the 1980s and 1990s, there was a renewed popular interest in Faraday, possibly stimulated by Margaret Thatcher becoming Prime Minister. She announced in a 1987 television programme that Faraday was her hero and suggested that his example showed that one could become a successful scientist without attending university(!). The following year in a speech to the Royal Society (who had elected her a Fellow in 1983, despite protests from a significant proportion of the Fellowship), she appreciated the monetary value of Faraday's work, which, she said, must exceed the capitalization of the Stock Exchange. Despite such admiration, the government she led pursued a policy of reducing both the science and university budgets. In response, Oxford University, braver than the Royal Society, refused to award her an honorary degree in 1985 by more than a two to one majority.

One response by the scientific community to government cuts was establishing in 1985 a joint committee (of the Royal Institution, Royal Society, and British Association) on the public understanding of science, to increase public awareness of science with the (deliberately unstated) aim of persuading the government to reverse its funding policies. One initiative proposed by the committee was founding an annual award to a practising scientist for significant contributions to the public understanding of science. Named the Faraday Award, the Royal Society added it to its portfolio of prizes. Won by scientists including Susan Greenfield, David Attenborough, George Porter, and Richard Dawkins, there is a strong bias towards making the award to Christmas Lecturers at the Royal Institution. The promotion of Faraday in the 1980s concluded with the unveiling by the Duke of Kent of a bronze copy of Foley and Brock's statue. Located near Waterloo Bridge, this is one of the few outdoor statues in London devoted to a scientific figure.

The major event in the following decade that brought Faraday before the general public was the bicentenary of his birth in 1991. (The 1967 centenary commemorations of his death had been muted, although the Royal Institution launched a major fundraising appeal linked to Faraday and a special BBC2 *Horizon* docudrama, staring the wonderful Ian Richardson, was televised.) For 1991, most importantly, the Bank of England replaced William Shakespeare on the £20 note with Faraday and Blaikley's picture of his Christmas Lecture. One cannot help but suspect that Thatcher gave backing for this note, which drew complaints from a number of people, epitomized by Peter Brookes's cartoon in *The Times* depicting, on a faux banknote, Faraday in his laboratory, scratching his chin and, in Hamlet mode, holding Shakespeare's head. The bicentenary events were quite diverse, though not on the scale of 1931. They lacked the coherence of the earlier celebrations, in terms of both organization and content. They included an excellent exhibition at the National Portrait Gallery, sponsored by the National Grid, and an exhibition at the Science Museum, sponsored by London Electricity. The electrical industry was then being denationalized, and these sponsorships suggest that the new companies realized the value of being associated with Faraday to promote their identities – once again, linking past and present.

Conclusion

To a greater or lesser extent, Faraday has been part of the general cultural landscape since the 1830s and is likely to remain so. Some recent examples illustrate this. Melvyn Bragg chose Faraday's 'Experimental Researches' as one of his *Twelve Books that Changed the World* (2006), and following the Royal Institution's reopening in 2008 after extensive refurbishment Bragg declared its Faraday Museum to be the 'most extraordinary museum in London'. The government programme established in 2006 to refurbish school science laboratories is called 'Project Faraday', and during the Parliamentary crisis of 2009, *The Daily Telegraph* republished the *Punch* cartoon (Figure 18) showing Faraday holding his nose at the stink of the Thames, surrounded by letters condemning the corrupt practices of MPs.

Part of the scientific landscape for a bit longer, Faraday is one of the figures most admired by modern scientists and engineers. During the expansion of the university system in the 1960s, buildings were named after him in universities with which he had no obvious connection, including Lancaster, Manchester, Salford, and Southampton. What this suggests is that those charged with selecting names believed that Faraday would carry greater recognition than most former vice-chancellors.

Would Faraday have been surprised by all the recognition, fame, and celebrity accorded him? He knew that his contributions to knowledge of the natural world were significant in terms of the development of science as he understood it. By the 1860s, it was clear that electricity would have a significant technological role in the future. Yet, despite his fondness for new technologies, such as railways and Brunel's large steamships, Faraday had no inkling of the dominant position that electricity occupies in modern engineering. But to fully answer the question is impossible. In all history, the historian eventually realizes that they cannot actually get into the head of their subject, cannot live their life for them – that is the preserve of the novelist (and those who have tried this approach to Faraday have produced remarkably unfruitful results).

In Faraday's case, the answer is further complicated by his religious beliefs, which required him to behave in ways significantly different from the norm of the scientific community and indeed the broader world. He would have known sitting in the pew or on the Elder's bench that to the world outside the Sandemanians, it would seem incongruous, at the very least, that here sat one of the most famous men in Europe as a member of one of the smallest and, in his word, 'despised' sects in the world. This must have produced a significant tension within him which can occasionally be glimpsed directly. A revealing exception to his usual self-control, patience, and diplomatic skill in resolving difficult situations was his walking out of the meeting with Melbourne in 1836. That incident showed a temporary loss of self-control, and indeed temper. Tyndall, who would have known, commented that 'Underneath his sweetness and gentleness was the heat of a volcano.'

Such incidents and descriptions do suggest the existence of a tension within Faraday which may have contributed to his creativity and obsessiveness. Furthermore, one cannot dismiss the huge number of portraits that exist of him simply because of his interest in art or the new photographic technology; they represent someone who wanted to tell the world he existed, but could not do so directly. Finally, when in 1866 George Barnard purchased Faraday's plot in Highgate Cemetery, he acquired two. One was for Faraday and Sarah and the other for a large hipstone (Figure 23) which dwarfs the surrounding Sandemanian graves.

It is not known whether Faraday authorized this – but Sarah must certainly have done so, but then she tended to be more ambitious for Faraday than he was. However it happened, it does look like Faraday trying to have it both ways: to be famous without appearing to desire fame.

But the final words should go to Faraday. In his letter to Dumas thanking him for his efforts in getting the Académie des Sciences to elect him one of their eight foreign members, the supreme accolade of 19th-century science, he reported: 'Mrs. Faraday…though she thinks very well of me still wonders to find that I can create such an interest in you and the Academy. She rejoices in my joy.' Faraday,

23. Laying of wreaths on Faraday's grave, 20 September 1931

clearly not indifferent to fame, in a letter to the famous organic chemist Justus Liebig, who had complained about the lack recognition accorded to his work in England, commented in regard to Liebig, what could equally have applied to him: 'how can a man expect, in his own life time, to be truly, recognized; it requires more than one generation to give currency to his highest truths'.

References and further reading

This gives references for the direct quotations used in the text. The first few words of each quotation are given, followed by its source. Following a short list of general books about Faraday, further reading is divided into chapters of the text, though, of course, some items may relate to more than one chapter.

Abbreviations

Printed sources

Correspondence	Frank A. J. L. James (ed.), *The Correspondence of Michael Faraday*, 5 vols and continuing (London, 1991–) (followed by volume and letter number). Letters to be published in sixth and final volume are referred to by author, recipient, and date.
Diary	Thomas Martin (ed.), *Faraday's Diary: Being the various philosophical notes of experimental investigation made by Michael Faraday, DCL, FRS, during the years 1820–1862 and bequeathed by him to the Royal Institution of Great Britain*, 7 vols and index (London, 1932–6; reprinted Riverton, 2008) (followed by date, volume, and paragraph number unless otherwise indicated).
ERE	Michael Faraday, *Experimental Researches in Electricity* (followed by series and paragraph number).

HBJ	Henry Bence Jones, *Life and Letters of Faraday*, 1st edn., 2 vols (London, 1870). (The second edition, published also in 1870, has slightly different pagination, especially in volume 1.)
JHG	John Hall Gladstone, *Michael Faraday* (London, 1872).
JT	John Tyndall, *Faraday as a Discoverer* (London, 1868).
SPT	Silvanus P. Thompson, *Michael Faraday: His Life and Work* (London, 1898).
T&G	Ryan Tweney and David Gooding (eds.), *Michael Faraday's 'Chemical Notes, Hints, Suggestions and Objects of Pursuit' of 1822* (London, 1991).

Manuscript sources

IET MS	Papers held by the Institution of Engineering and Technology (formerly the Institution of Electrical Engineers).
RI MS F4	Faraday's lecture notes held in the Royal Institution.
RI MS HD	Papers of Humphry Davy held in the Royal Institution.
RI MS MM	Minutes of the meetings of Managers of the Royal Institution, followed by date of meeting, volume and page number. Volumes 1 to 15 were published in facsimile in seven bound volumes in 1971–6.
RI MS Pep	Papers of William Pepys held in the Royal Institution.
RS MS 241	Faraday's diploma book held in the Royal Society.
ULC add MS 7342	Papers of William Thomson (Lord Kelvin) held in the University Library, Cambridge.

References

Introduction

'for the good of Australia', *Correspondence* 2: 1031.

'that true old English Gentleman', *Correspondence* 3: 1942.

Chapter 1

'common day-school' and 'my education was...', HBJ 1: 9.

'When they first formed the canal at Paddington...', Faraday to
  George Buchanan, 23 December 1862.

'He could remember none', JT 3.

'distrait' and 'the noise reminded me of my father's anvil...',
  JHG 79.

'separatist', HBJ 1: 1.

'that horrid blast from the north', Luke Tyerman, *The Oxford
  Methodists: Memoirs of the Rev. Messrs. Clayton, Ingham,
  Gambold, Hervey, and Broughton, with biographical notices of
  others* (London, 1873), p. 140.

'a pillow to my mind', Faraday to Sarah Faraday, 14 August 1863.

'There was no eloquence', HBJ 2: 100.

'no ceremony', Highgate Cemetery MS Register of Burials,
  number 2839.

'if, in his deepest conviction...', JHG 91.

'we are then all three bound for the one great object', Elizabeth Mary
  Odling, *Memoir of the Late Alfred Smee* (London, 1878), p. 61.

'Printed for George Riebau...', Richard Brothers, *A Description
  of Jerusalem: Its houses and streets, squares, colleges, markets, and
  cathedrals, the royal and private palaces, with the Garden of Eden
  in the centre as laid down in the last chapters of Ezekiel. Also the
  first chapter of Genesis verified, as strictly divine and true. And the
  solar system, with all its plurality of inhabited worlds, and millions
  of suns, as positively proved to be delusive and false. By Mr.
  Brothers, who will be revealed to the Hebrews as their king and
  restorer* (London, 1801), title page.

'I always feel a tenderness...', HBJ 1: 10–11.

'the Art of Bookbinding...' and 'in consideration of his...',
  RS MS 241, f.1.

'a very good master...', HBJ 1: 11.

'in his younger days...' and 'perhaps to learn...', SPT 290.

'I was a very lively...', 'in the hours after work', and 'I could trust a
  fact...', *Correspondence* 5: 3519.

'miniature Laboratory', Frank A. J. L. James, 'The Tales of Benjamin Abbott: A Source for the Early Life of Michael Faraday', *British Journal for the History of Science*, 1992, 25: 229–40, p. 236.

'vicious and selfish' and 'to enter into the service of Science', *Correspondence* 1: 419.

'much trouble', HBJ 1: 46.

'received a Guinea & half…', *Correspondence* 1: 30.

'to leave at the first convenient opportunity', *Correspondence* 1: 15.

'grateful for the goodness of Mr. de la Roche', *Correspondence* 1: 35.

Chapter 2

'attended twelve or thirteen lectures…', and 'allowed', HBJ 1: 11–12.

'Institution For diffusing the Knowledge…', RI MS MM, 9 March 1799, 1: 1.

'Newton & Davy', RI MS HD 20b, p. 182.

'Had not Davy been the first chemist…', John Davy, *Fragmentary Remains, Literary and Scientific, of Sir Humphry Davy* (London, 1858), pp. 322–3.

'Alas Alas Inability', *Correspondence* 1: 3.

'engaged in scientific occupation…' and 'no answer', HBJ 1: 14.

'The reply was immediate, kind, and favourable', HBJ 1: 53.

'great zeal power of memory & attention', *Correspondence* 1: 17.

'quite recovered', Davy to John Davy, 4 April 1813, in John Davy, *Memoirs of the Life of Sir Humphry Davy*, 2 vols (London, 1836), 1: 459–60 on p. 459.

'Attend to the book binding', *Correspondence* 1: 30.

'great noise' and 'at high words', RI MS MM, 22 February 1813, 5: 353.

'Sir Humphry Davy has the honor…', RI MS MM, 1 March 1813, 5: 355.

'Science was harsh mistress' and 'notion of the superior moral feelings…', *Correspondence* 1: 419.

'I have been engaged this afternoon…', *Correspondence* 1: 19.

'philosophical assistant', *Correspondence* 1: 28.

'What an extraordinary collection of fine frames', John Ayrton Paris, *The Life of Sir Humphry Davy* (London, 1831), p. 268.

'we owe to the sagacity of Dr. Wollaston…', 'Proceedings of Philosophical Societies. Royal Society', *Annals of Philosophy*, 1823, 5: 300–5, p. 304.

'Then he said, I as President…', HBJ 1: 379.

'by no means in the same relation…', HBJ 1: 353.

'would not answer for the integrity…', JT 157–8.

Chapter 3

'three or four philosophical discoveries', *Correspondence* 2: 606.

'We trouble you as a universal referee...', *Correspondence* 3: 1990.

'correct in theory...', *Correspondence* 4: 2871.

'accident' and 'verdict would be delivered...', *Correspondence* 3: 1616.

'Faraday began, after a few minutes...', Lyell to Bence Jones, April 1868, HBJ 2: 186.

'He sprung up on his feet...', HBJ 2: 186.

'fully agree with them', HBJ 2: 185.

'Scientific adviser in experiments on Lights to the Corporation', *Correspondence* 2: 884.

'by climbing over hedges, walls, and fields', *Correspondence* 5: 3728.

'The use of light to guide the mariner...', Michael Faraday, 'On Lighthouse Illumination – the Electric Light', *Proceedings of the Royal Institution*, 1860, 3: 220–3, on p. 220.

'Much, therefore, as I desire to see the Electric light...', *Correspondence* 4: 2878.

'conveys the gifts of God to Man', Michael Faraday, 'On Wheatstone's Electric Telegraph in Relation to Science (being an argument in favour of the full recognition of Science as a branch of Education)', *Proceedings of the Royal Institution*, 1858, 3: 555–60, on p. 560.

Chapter 4

'Convert magnetism into Electricity', T&G, 70–1.

'Expts. On the production of Electricity from Magnetism, etc etc', *Diary*, 29 August 1831, 1: 1.

'I have got hold of a good thing...', *Correspondence* 1: 515.

'for health's sake', *Correspondence* 1: 521.

'or else these philosophers...', *Correspondence* 2: 531.

'I never took more pains...', *Correspondence* 2: 560.

'that magnetic action is progressive, and requires time', *Correspondence* 2: 557.

'as an axis of power...', ERE 5, 517.

'produced by an internal corpuscular action', ERE 5, 518.

'wanted some new names to express my facts in Electrical science', *Correspondence* 2: 711.

'that names are one things and science another', ERE 7, 666.

'almost as anti-material a view as myself...', Hamilton to Sydney Hamilton, 30 June 1834, in R. P. Graves, *Life of Sir William Rowan Hamilton*, 3 vols (Dublin, 1882–91), 2: 95–6, on p. 96.

'went into the cube and lived in it' and 'lighted candles,
  electrometers...', ERE 11, 1174.

'view jumps in with my notion...', *Correspondence* 2: 954.

'value and probable success', William Whewell, *History of the Inductive
  Sciences, From the Earliest to the Present Times*, 3 vols (London,
  1837), 3: 39.

'a connecting link between conduction and induction', *Diary*, 3
  November 1835, 2: 2554.

'in peace & quietness', *Correspondence* 2: 1045.

'common induction was in all cases...', ERE 11, 1164.

'the instruments of life', [Mary Shelley], *Frankenstein; or, the Modern
  Prometheus*, 3 vols (London, 1818), 1: 97.

'The New Frankenstein', 'The New Frankenstein', *Fraser's Magazine*,
  1838, 17: 21–30.

'Wonderful as are the laws and phenomena...' and 'We are indeed but
  upon the threshold...', ERE 15, 1749.

'observed that every remark he might make...', *Literary Gazette*,
  26 January 1839, p. 58.

'scarcely likely', *Diary*, 9 November 1835, 2: 2613.

'Compare corpuscular forces...', *Diary*, 6 November 1837, 3: 4216.

'ALL THIS IS A DREAM...', *Diary*, 19 March 1849, 5: 10040.

'The results are negative...', ERE 24: 2717.

Chapter 5

'Matter and space query their nature...', RI MS F4 G13.

'matter is not merely mutually penetrable...' and 'seems to fall in very
harmoniously...', Michael Faraday, 'A speculation touching Electric
  Conduction and the Nature of Matter', *Philosophical Magazine*,
  1844, 24: 136–44, on p. 143.

'I have been so long delayed...', *Correspondence* 3: 1642.

'little hope', *Correspondence* 3: 1663.

'I have been very much disgusted...', ULC add MS 7342, NB29.

'thus magnetic force and light...', *Diary*, 13 September 1845, 4: 7504.

'I found I *could* affect it...', *Diary*, 4 November 1845, 4: 7902.

'You can hardly imagine...', *Correspondence* 3: 1785.

'He who proves discovers', Herschel to Pepys, 9 November 1845, RI MS
  Pep A40.

'*matter* only by its powers', ERE 19, 2225.

'The view which I am so bold as to put forth...' and 'The smallest atom
  of matter...', Michael Faraday, 'Thoughts on Ray-vibrations',
  Philosophical Magazine, 1846, 28: 345–50, pp. 348 and 346.

'failed to observe the effect years ago', Michael Faraday, 'On the Diamagnetic Conditions of Flame and Gases', *Philosophical Magazine*, 1847, 31: 401–21, p. 401.

'Perhaps the cause here...', *Diary*, 23 July 1850, 5: 10958.

'keep this to yourself', *Correspondence* 4: 2317.

'works out beautifully', *Correspondence* 4: 2322.

'I was at first almost frightened, *Correspondence* 5: 3620.

'[Maxwell] wrote a paper...', *Correspondence* 5: 3364.

'not be expressed in common language...', *Correspondence* 5: 3357.

'The conception of the propagation...', J. C. Maxwell, 'A Dynamical Theory of the Electro-Magnetic Field', *Philosophical Transactions*, 1865, 155: 459–512, on p. 466.

'the nucleus of everything electric since 1830', Maxwell to Litchfield, 7 February 1858, in P. M. Harman (ed.), *The Scientific Letters and Papers of James Clerk Maxwell, Volume 1. 1846-1862* (Cambridge, 1990), p. 582.

'who took in Faraday's ideas...', Albert Einstein, 'Considerations Concerning the Fundaments of Theoretical Physics', *Science*, 1940, 91: 487–92, p. 488.

'the electric field theory...', Albert Einstein, 'Physics and Reality', *Journal of the Franklin Institute*, 1936, 221: 349–82, p. 363.

Chapter 6

'[It was] proposed to take them...', *Correspondence* 5: 3541.

'the *beau idéal* of a popular lecturer', Richard Owen, *The Life of Richard Owen*, 2 vols (London, 1894), 1: 153.

'as fashionable an amusement as the Opera', George Eliot to Charles and Caroline Bray, 28 January 1851, in Gordon S. Haight, *The George Eliot Letters, Volume 1, 1836-1851* (New Haven, 1954), 341–4, on p. 341.

'humbug' and 'theological', IET MS SC 2/3/18/10.

'whatever that may signify', 'Gallery of literary characters. No. LXIX. Michael Faraday, F.R.S., Hon.D.C.L. Oxon, Etc. Etc.', *Fraser's Magazine*, 1836, 13: 224.

'not disposed to place faith...', Walter White, *The Journals of Walter White* (London, 1898), p. 69.

'On perusing the 21st Series...', *Correspondence* 3: 1984.

'Mesmeric stuff', *Correspondence* 3: 1892.

'numerous', 'a *quasi* involuntary muscular action' and 'I think the system of education...', *Correspondence* 4: 2691.

'arrogance & insolence', Browning to Ogilvy, 21 July 1853, in
    P. N. Heydon and P. Kelley, *Elizabeth Barrett Browning's Letters to*
    Mrs. David Ogilvy 1849–1861 (London, 1974), pp. 100–4, on p. 102.
'shocked at the flood of impious…', *Correspondence* 4: 2703.
'keenly criticised', *Correspondence* 4: 2819.
'We remember him more as a man…', Faraday to Ernst Becker,
    1 January 1862.
'No human hand has hitherto traced…', 'Royal Institution',
    Literary Gazette, 2 February 1839, pp. 74–5, on p. 75.
'whether all persons are in this curious condition', Faraday to
    John Watkins, 9 April 1861.
'pupil', *Art Journal*, 1851, 13: 30.
'dark sayings', JT 72.

Chapter 7

'admirable and lucid way in which he delivered', *Proceedings of the*
    Royal Institution, 1891, 13: 462.
'Communism is Soviet power plus…', quoted in Jonathan Cooper-
    smith, 'Technology Transfer in Russian Electrification, 1870–1825',
    History of Technology, 1991, 13: 214–33, p. 221.
'one of the earliest Gas Consultants', RI MS uncatalogued.
'the question of disposing of…', Minutes of IEE Council, 27 March
    1930, IET MS ORG/2/1/14, p. 52.
'great Commemorative Meeting', Minutes of IEE Faraday Celebrations
    Committee, 3 October 1929, IET MS ORG/4/2/5, p. 88.
'Without Michael Faraday there would have been no broadcast' and
'What a miserable personality that has no contradictions!', 'Faraday
    Celebrations, 1931', *Proceedings of the Royal Institution*, 1932, 27:
    1–72, pp. 20–3.
'expressed the view that…', RI MS MM, 17 November 1932, 20: 136.
'the importance of the system…', 'The life of Faraday becomes a film',
    Picture Post, 25 July 1942, pp. 16–17 on p. 17.
'most extraordinary museum in London', 'In Our Time',
    BBC Radio 4, 12 November 2009.
'despised', *Correspondence* 3: 1631.
'Underneath his sweetness…', JT 37.
'Mrs. Faraday… though she thinks…', *Correspondence* 3: 1663.
'how can a man expect…', *Correspondence* 5: 192.

Further reading

General

Geoffrey Cantor, *Michael Faraday: Sandemanian and Scientist, A Study of Science and Religion in the Nineteenth Century* (London, 1991).

David Gooding, *Experiment and the Making of Meaning: Human Agency in Scientific Observation and Experiment* (Dordrecht, 1991).

David Gooding and Frank A. J. L. James (eds.), *Faraday Rediscovered: Essays on the Life and Work of Michael Faraday, 1791-1867* (London, 1985).

Frank A. J. L. James (ed.), *'The Common Purposes of Life': Science and Society at the Royal Institution of Great Britain* (Aldershot, 2002).

Alan E. Jeffreys, *Michael Faraday: A List of His Lectures and Published Writings* (London, 1960).

Manuscripts

The vast majority of Faraday's manuscripts, apart from letters, have been published on microfilm and CD. Frank A. J. L. James, *Guide to the Microfilm Edition of the Manuscripts of Michael Faraday (1791-1867) from the Collections of the Royal Institution, The Institution of Electrical Engineers, The Guildhall Library [and] The Royal Society*, 2nd edn. (Wakefield, 2001).

Introduction

For overviews of Faraday's scientific approach, see David Gooding, 'Empiricism in Practice: Teleology, Economy, and Observation in Faraday's Physics', *ISIS*, 1982, 73: 46–67; Ryan Tweney, 'Fields of Enterprise: On Michael Faraday's Thought', in *Creative People at Work: Twelve Cognitive Case Studies*, ed. Doris B. Wallace and Howard E. Gruber (New York, 1989), pp. 91–106; and Ronald Anderson, 'The Crafting of Scientific Meaning and Identity: Exploring the Performative Dimensions of Michael Faraday's Texts', *Perspectives on Science*, 2006, 14: 7–39. For Faraday and mathematics, see David Gooding, 'Mathematics and Method in Faraday's Experiments', *Physis*, 1992, 29: 121–47; and for his note-taking, see Ryan Tweney, 'Faraday's Notebooks: The Active Organization of Creative Science', *Physics*

Education, 1991, 26: 301–6. On Faraday's *Chemical Manipulation*, see William B. Jensen, 'Michael Faraday and the Art and Science of Chemical Manipulation'; and Sydney Ross, 'The Chemical Manipulator', *Bulletin for the History of Chemistry*, 1991, 11: 65–76 and 76–9, respectively. For the Banksian programme, see John Gascoigne, *Science in the Service of Empire: Joseph Banks, the British State and the Uses of Science in the Age of Revolution* (Cambridge, 1998).

Chapter 1

For the Faraday family's economic and demographic background, see Phyllis Deane and W. A. Cole, *British Economic Growth 1688–1959* (Cambridge, 1967) and, more specifically, James Frederic Riley, *The Hammer and the Anvil: A Background to Michael Faraday* (Clapham, 1954), which also covers the Sandemanians, but contains inaccuracies. For the Sandemanians generally, see John Howard Smith, *'The Prefect Rule of the Christian Religion': A History of Sandemanianism in the Eighteenth Century* (Albany, 2008). For Faraday and the Sandemanians, see Geoffrey Cantor, *Michael Faraday*, and, for Ada Lovelace, his 'Michael Faraday Meets the 'High-Priestess of God's Works': A Romance on the Theme of Science and Religion', in *Science and Beliefs: From Natural Philosophy to Natural Science, 1700–1900*, ed. Matthew Eddy and David Knight (Aldershot, 2005), pp. 157–70. On the Barnards and their business, see John Fallon, 'The House of Barnard', *Silver Studies*, 2009, 25: 42–62. See also Herbert T. Pratt, 'Michael Faraday's Bible as Mirrors of His Faith', *Bulletin for the History of Chemistry*, 1991, 11: 40–7. For a general overview, see Thomas Dixon, *Science and Religion: A Very Short Introduction* (Oxford, 2008).

On the millenarians, see J. F. C. Harrison, *The Second Coming: Popular Millenarianism, 1780–1850* (London, 1979) and for further, but limited, information on bookbinding, Charles Ramsden, *London Bookbinders, 1780–1840* (London, 1987). For Faraday's essay circle, see Alice Jenkins, *Michael Faraday's Mental Exercises: An Artisan Essay Circle in Regency London* (Liverpool, 2008).

Chapter 2

On Tatum and the City Philosophical Society, see Frank A. J. L. James, 'Michael Faraday, The City Philosophical Society and the Society of Arts', *Royal Society of Arts Journal*, 1992, 140: 192–9.

For the Royal Institution, in addition to the essays by Sophie Forgan in Gooding and James, *Faraday Rediscovered* and James, '*Common Purposes of Life*', see Frank A. J. L. James and Anthony Peers, 'Constructing Space for Science at the Royal Institution of Great Britain', *Physics in Perspective*, 2007, 9: 130–85. Morris Berman, *Social Change and Scientific Organization: The Royal Institution, 1799-1844* (London, 1978) contains a wealth of information on the early Royal Institution, but interpretively is now dated. For the Smithsonian Institution, see Heather Ewing, *The Lost World of James Smithson: Science, Revolution, and the Birth of the Smithsonian* (London, 2007).

On Davy, see David Knight, *Humphry Davy: Science and Power* (Oxford, 1992; 2nd edn., Cambridge, 1996) as well as his essays in Gooding and James, *Faraday Rediscovered* and James, '*Common Purposes of Life*'. See also Sophie Forgan (ed.), *Science and the Sons of Genius: Studies on Humphry Davy* (London, 1980). More generally, see Jan Golinski, *Science as Public Culture: Chemistry and Enlightenment in Britain, 1760-1820* (Cambridge, 1992).

Those parts of Faraday's Continental diary which have survived (mostly from the first half of the journey) have been published in Brian Bowers and Lenore Symons, '*Curiosity Perfectly Satisfyed': Faraday's Travels in Europe 1813-1815* (London, 1991).

On the miners' safety lamp, see Frank A. J. L. James, 'How Big is a Hole?: The Problems of the Practical Application of Science in the Invention of the Miners' Safety Lamp by Humphry Davy and George Stephenson in Late Regency England', *Transactions of the Newcomen Society*, 2005, 75: 175–227. On the Royal Society generally, see Marie Boas Hall, *All Scientists Now: The Royal Society in the Nineteenth Century* (Cambridge, 1984); and for Davy's role, L. F. Gilbert, 'The Election to the Presidency of the Royal Society in 1820', *Notes and Records of the Royal Society of London*, 1955, 11: 256–79, and David Philip Miller, 'Between Hostile Camps: Sir Humphry Davy's Presidency of the Royal Society of London 1820-1827', *British Journal for the History of Science*, 1983, 16: 1–47. For Faraday's election, June Z. Fullmer and Melvyn C. Usselman, 'Faraday's Election to the Royal Society: A Reputation in Jeopardy', *Bulletin for the History of Chemistry*, 1991, 11: 17–28. For Faraday's metallurgical work, see Robert A. Hadfield, *Faraday and His Metallurgical Researches, with special reference to their bearing on the development of alloy steels* (London, 1931). For Faraday's discovery of electro-magnetic rotations, see Gooding, *Experiment and the Making of Meaning* and his essay in Gooding and James, *Faraday Rediscovered*, as well as his 'Experiment and Concept Formation in Electromagnetic Science and Technology in England in the 1820s', *History and Technology*, 1985, 2: 151–76; "Magnetic Curves" and the Magnetic Field: Experimentation and Representation in the History of a Theory', in *The Uses of Experiment: Studies in the Natural Sciences*, ed. David Gooding, Trevor Pinch, and Simon Schaffer (Cambridge, 1989), pp. 183–223; 'History in the Laboratory: Can We Tell What Really Went On?', in *The Development of the Laboratory: Essays on the Place of Experiment in Industrial Civilisation*, ed. Frank A. J. L. James (London, 1989), pp. 63–82; and 'Mapping Experiment as a Learning Process: How the First Electromagnetic Motor Was Invented', *Science Technology and Human Values*, 1990, 15: 165–201. See also Friedrich Steinle, 'Looking for a 'Simple Case': Faraday and Electromagnetic Rotation', *History of Science*, 1995, 33: 179–202.

Chapter 3

For the protection of naval vessels, see Frank A. J. L. James, 'Davy in the Dockyard: Humphry Davy, the Royal Society and the Electro-Chemical Protection of the Copper Sheeting of His Majesty's Ships in the mid 1820s', *Physis*, 1992, 29: 205–25; for the glass work, Frank A. J. L. James, 'Michael Faraday's Work on Optical Glass', *Physics Education*, 1991, 26: 296–300; and for the Royal Military Academy, his 'The Military Context of Chemistry: The Case of Michael Faraday', *Bulletin for the History of Chemistry*, 1991, 11: 36–40.

For Haswell colliery, see Frank A. J. L. James and Margaret Ray, 'Science in the Pits: Michael Faraday, Charles Lyell and the Home Office Enquiry into the Explosion at Haswell Colliery, County Durham, in 1844', *History and Technology*, 1999, 15: 213–31. On the war against Russia, see Andrew Lambert, *The Crimean War: British Grand Strategy, 1853–56* (Manchester, 1990).

For Faraday's lighthouse work, see Frank A. J. L. James, '"the Civil-Engineer's Talent": Michael Faraday, Science, Engineering and the English Lighthouse Service, 1836–1865', *Transactions of the Newcomen Society*, 1999: 70: 153–60, and 'Michael Faraday and Lighthouses' in Ian Inkster, Colin Griffin, Jeff Hill, and Judith Rowbotham (eds.), *The Golden Age: Essays in British Social and Economic History, 1850–1870* (Aldershot, 2000), pp. 92–104. For contemporary comments on electrification, see J. J. W. Watson,

A Few Remarks on the Present State and Prospects of Electrical Illumination (London, 1853) and F. H. Holmes, *Holmes's Magneto-Electric Light, as applicable to Lighthouses* (London, 1862). The report of the Royal Commission on Lighthouses is in *Parliamentary Papers*, 1861 [2793], XXV.

Chapter 4

The best recent accounts of Faraday's discovery of electro-magnetic induction are Ryan Tweney's essay in Gooding and James, *Faraday Rediscovered*, and chapter 5 of Albert E. Moyer, *Joseph Henry: The Rise of an American Scientist* (Washington, 1997). For Arago's experiment, see Friedrich Steinle, 'Experiment, Speculation and Law: Faraday's Analysis of Arago's Wheel', *PSA: Proceedings of the Biennial Meeting of the Philosophy of Science Association*, 1994, 1: 293–303. Also valuable are Sydney Ross, 'The Search for Electromagnetic Induction', *Notes and Records of the Royal Society of London*, 1965, 20: 184–219; José Romo and Manuel G. Doncel, 'Faraday's Initial Mistake Concerning the Direction of Induced Currents, and the Manuscript of Series I of His *Researches*', *Archive for the History of the Exact Sciences*, 1994, 47: 291–385; Ronald Anderson, 'The Referees' Assessment of Faraday's Electromagnetic Induction Paper of 1831', *Notes and Records of the Royal Society of London*, 1993, 47: 243–56; Friedrich Steinle, 'The Practice of Studying Practice: Analyzing Laboratory Records of Ampère and Faraday', in *Reworking the Bench: Research Notebooks in the History of Science*, ed. Frederic Lawrence Holmes, Jürgen Renn, and Hans-Jörg Rheinberger (Dordrecht, 2003), pp. 93–118, especially pp. 106–13, and his 'Work, Finish, Publish?: The Formation of the Second Series of Faraday's Experimental Researches in Electricity', *Physis*, 1996, 33: 141–220. For the technicalities of Faraday's discovery, see Elizabeth Cavicchi, 'Nineteenth-Century Developments in Coiled Instruments and Experiences with Electromagnetic Induction', *Annals of Science*, 2006, 63: 319–61; Allan A. Mills, 'The Early History of Insulated Copper Wire', *Annals of Science*, 2004, 61: 453–67; and B. C. Blake-Coleman and R. Yorke, 'Faraday and Electrical Conductors: An Examination of the Copper Wire Used by Michael Faraday between 1821 and 1831', *Proceedings of the Institution of Electrical Engineers*, 1981, 128A: 463–71. For the Nobili incident, see Pasquale Tucci, 'Faraday contro Nobili: un episodio della polemica antiampèriana', *Giornale di fisica della Società italiana di fisica*, 1984, 25: 347–64.

For Faraday and electro-chemistry, see Sydney Ross, 'Faraday Consults the Scholars: The Origins of the Terms of Electrochemistry', *Notes and Records of the Royal Society of London*, 1961, 16: 187–220; Stanley M. Guralnick, 'The Contexts of Faraday's Electrochemical Laws', *ISIS*, 1979, 70: 59–75; Frank A. J. L. James, 'Michael Faraday's First Law of Electrochemistry: How Context Develops New Knowledge', in *Electrochemistry, Past and Present*, ed. John T. Stock and Mary Virginia Orna (Washington, 1989), pp. 32–49; John T. Stock, 'The Pathway to the Laws of Electrolysis', *Bulletin for the History of Chemistry*, 1991, 11: 86–92. See also Simon Schaffer, 'The History and Geography of the Intellectual World: Whewell's Politics of Language', in *William Whewell: A Composite Portrait*, ed. Menachem Fisch and Simon Schaffer (Oxford, 1991), pp. 201–31.

For Faraday's work on fluorine, see Harold Goldwhite, 'Faraday's Search for Fluorine', *Bulletin for the History of Chemistry*, 1991, 11: 55–60. For the Faraday cage and electric discharge, see David Gooding's essay in Gooding and James, *Faraday Rediscovered* and his 'Conceptual and Experimental Bases of Faraday's Denial of Electrostatic Action at a Distance', *Studies in the History and Philosophy of Science*, 1978, 9: 117–49. For Faraday's role in developing spectroscopy, see Frank A. J. L. James, 'The Study of Spark Spectra, 1835–1859', *Ambix*, 1983, 30: 137–62.

There is an enormous literature on Frankenstein, but for an introduction to the role of contemporary science, see J. V. Field and Frank A. J. L. James, 'Frankenstein and the Spark of Being', *History Today*, September 1994, pp. 47–53. On Crosse's insects, see James A. Secord, 'Extraordinary Experiment: Electricity and the Creation of Life in Victorian England', in *The Uses of Experiment: Studies in the Natural Sciences*, ed. David Gooding, Trevor Pinch, and Simon Schaffer (Cambridge, 1989), pp. 337–83, and Oliver Stallybrass, 'How Faraday 'Produced Living Animalculae': Andrew Crosse and the Story of a Myth', *Proceedings of the Royal Institution*, 1967, 41: 597–619. For electric fish, see Howard Fisher, 'The Body Electric', *St John's Review*, 1991–2, 41: 1–37.

On Faraday's gravitational work, see Geoffrey Cantor, 'Faraday's Search for the Gravelectric Effect', *Physics Education*, 1991, 26: 289–93, and section 9.2 of his *Michael Faraday*. See also Frans van Lunteren, *Framing Hypotheses: Conceptions of Gravity in the 18th and 19th Centuries* (Utrecht, 1991). For Faraday's views on conservation, see David Gooding, 'Metaphysics versus

Measurement: The Conversion and Conservation of Force in Faraday's Physics', *Annals of Science*, 1980, 37: 1–29.

Chapter 5

For Faraday's health, see Edward Hare, 'Michael Faraday's Loss of Memory', *Proceedings of the Royal Institution*, 1976, 49: 33–52, and for his problem situation in the 1840s, Frank James's essay in Gooding and James, *Faraday Rediscovered*. On the magneto-optical effect, see David Gooding, "He Who Proves Discovers': John Herschel, William Pepys and the Faraday Effect', *Notes and Records of the Royal Society of London*, 1985, 39: 229–44; J. Brooks Spencer, 'On the Varieties of Nineteenth-Century Magneto-Optical Discovery', *ISIS*, 1970, 61: 34–51; and Ole Knudsen, 'The Faraday Effect and Physical Theory, 1845–1873', *Archive for History of the Exact Sciences*, 1976, 15: 235–8. On William Thomson (Lord Kelvin from 1892) and Faraday, see David Gooding, 'A Convergence of Opinion on the Divergence of Lines: Faraday and Thomson's Discussion of Diamagnetism', *Notes and Records of the Royal Society of London*, 1982, 36: 243–59; Jed Z. Buchwald, 'William Thomson and the Mathematization of Faraday's Electrostatics', *Historical Studies in the Physical Sciences*, 1977, 8: 101–36; and, more generally, M. Norton Wise and Crosbie Smith, *Energy and Empire: A Biographical Study of Lord Kelvin* (Cambridge, 1989).

On Bancalari, see Giovanni Boato and Natalia Moro, 'Bancalari's Role in Faraday's Discovery of Diamagnetism and the Successive Progress in the Understanding of Magnetic Properties of Matter', *Annals of Science*, 1994, 51: 391–412. On the establishment of magnetic observatories across the globe, see John Cawood, 'Terrestrial Magnetism and the Development of International Collaboration in the Early Nineteenth Century', *Annals of Science*, 1977, 34: 551–87, and 'The Magnetic Crusade: Science and Politics in Early Victorian Britain', *ISIS*, 1979, 70: 493–518.

On field theory, see David Gooding, 'Faraday, Thomson, and the Concept of the Magnetic Field', *British Journal for the History of Science*, 1980, 13: 91–120, 'Final Steps to the Field Theory: Faraday's Study of Magnetic Phenomena, 1845–1850', *Historical Studies in the Physical Sciences*, 1981, 11: 231–75, and 'From Phenomenology to Field Theory: Faraday's Visual Reasoning', *Perspectives on Science*, 2006, 14: 40–65. See also Ryan Tweney, 'Inventing the Field: Michael Faraday and the Creative 'Engineering' of Electromagnetic

Field Theory', in *Inventive Minds: Creativity in Technology*, ed. Robert J. Weber and David H. Perkins (New York, 1992), pp. 31–47. For the role of long-distance telegraphy, Gillian Cookson, *The Cable: The Wire That Changed the World* (Stroud, 2003); Bruce J. Hunt, 'Michael Faraday, Cable Telegraphy and the Rise of Field Theory', *History of Technology*, 1991, 13: 1–19, and his 'Insulation for an Empire: Gutta-Perhca and the Development of Electrical Measurement in Victorian Britain', in *Semaphores to Short Waves*, ed. Frank A. J. L. James (London, 1998), pp. 85–104. See also Frank A. J. L. James, 'Faraday, Maxwell and Field Theory', in *ibid.*, pp. 71–84; P. M. Harman, 'Maxwell and Faraday', *European Journal of Physics*, 1993, 14: 148–54; and, more generally, Bruce J. Hunt, *The Maxwellians* (Ithaca, 1991) and Nancy J. Nersessian, *Faraday to Einstein: Constructing Meaning in Scientific Theories* (Dordrecht, 1984). For Maxwell's view of Faraday, see his 'Scientific Worthies I. – Faraday', *Nature*, 1873, 8: 397–9.

Chapter 6

On Faraday's lecturing, see Frank A. J. L. James, 'Reporting Royal Institution Lectures, 1826 to 1867', in *Science Serialized: Representations of the Sciences in Nineteenth-Century Periodicals*, ed. Sally Shuttleworth and Geoffrey Cantor (Cambridge, MA, 2004), pp. 67–79; Geoffrey Cantor, 'How Michael Faraday Brought Law and Order to the West End of London', *Physis*, 1992, 29: 187–203, and his 'Educating the Judgment: Faraday as a Lecturer', *Bulletin for the History of Chemistry*, 1991, 11: 28–36; Frank A. J. L. James (ed.), 'Introduction' to *Christmas at the Royal Institution: An Anthology of Lectures by M. Faraday, J. Tyndall, R. S. Ball, S. P. Thompson, E. R. Lankester, W. H. Bragg, W. L. Bragg, R. L. Gregory, and I. Stewart* (Singapore, 2007), pp. xi–xxv. See also Isobel Falconer and Frank A. J. L. James, 'Fame and Faraday', in *Reputations*, ed. Elaine Moohan (Milton Keynes, 2008), pp. 85–122.

There is a large literature on mesmerism and spiritualism in 19th-century Britain, including Alison Winter, *Mesmerized: Powers of Mind in Victorian Britain* (Chicago, 1998); Janet Oppenheim, *The Other World: Spiritualism and Psychical Research in England, 1850–1914* (Cambridge, 1985); and W. H. Brock, *William Crookes (1832–1919) and the Commercialization of Science* (Aldershot, 2008). For Faraday and Prince Albert, see Frank A. J. L. James, *The Royal Institution and the Royal Family 1799–1999* (London, 1999). For

Faraday's lecture on mental education, see Elspeth Crawford's essay in Gooding and James, *Faraday Rediscovered*.

For detailed studies of Faraday and visual images, see G. M. Prescott's essays in Gooding and James, *Faraday Rediscovered* and James, *'Common Purposes of Life'*, and Frank A. J. L. James, 'Harriet Jane Moore, Michael Faraday, and Moore's Mid-Nineteenth Century Watercolours of the Interior of the Royal Institution', in *Fields of Influence: Conjunctions of Artists and Scientists, 1815–1860*, ed. James Hamilton (Birmingham, 2001), pp. 111–28. For an analysis of Faraday's early biographers, see Geoffrey Cantor, 'The Scientist as Hero: Public Images of Michael Faraday', in *Telling Lives in Science: Essays on Scientific Biography*, ed. M. Shortland and R. Yeo (Cambridge, 1996), pp. 171–93. For the background to photography generally, see Larry J. Schaaf, *Out of the Shadows: Herschel, Talbot and the Invention of Photography* (New Haven, 1992).

Chapter 7

On the electrical industry generally, L. Hannah, *Electricity before Nationalisation: A Study of the Development of the Electricity Supply Industry in Britain to 1948* (London, 1979); B. Luckin, *Questions of Power: Electricity and Environment in Inter-War Britain* (Manchester, 1990) (this contains a useful discussion of the Electrical Development Association); W. J. Reader, *A History of the Institution of Electrical Engineers, 1871–1971* (London, 1987); I. C. R. Byatt, *The British Electrical Industry 1875–1914: The Economic Returns to a New Technology* (Oxford, 1979). For Russian electrification, see Jonathan Coopersmith, *The Electrification of Russia, 1880–1926* (Ithaca, 1992). There is quite a large literature on the Electrical Association for Women, including Carroll Pursell, 'Domesticating Modernity: The Electrical Association for Women, 1924–86', *British Journal for the History of Science*, 1999, 32, 47–56. For Finsbury Technical College, see W. H. Brock, 'Building England's First Technical College: The Laboratories of Finsbury Technical College, 1878–1926', in *The Development of the Laboratory: Essays on the Place of Experiment in Industrial Civilisation*, ed. Frank A. J. L. James (London, 1989), pp. 155–70. For Faraday House, see Frank W. Lipscomb, *The Wise Men of the Wires: The History of Faraday House* (London, 1973), and for Faraday specifically, see Graeme Gooday, 'Faraday Reinvented: Moral Imagery and Institutional Icons in Victorian Electrical Engineering', *History of Technology*, 1993, 15: 190–205.

Frank A. J. L. James, 'Presidential Address. The Janus Face of Modernity: Michael Faraday in the Twentieth Century', *British Journal for the History of Science*, 2008, 41: 477–516.

For the use of Faraday in cognitive science, see Ryan Tweney's papers, including 'Toward a Cognitive-Historical Understanding of Michael Faraday's Research: Editor's Introduction', *Perspectives on Science*, 2006, 14: 1–6; 'Stopping Time: Faraday and the Scientific Creation of Perceptual Order', *Physis*, 1992, 29: 149–64; 'Procedural Representation in Michael Faraday's Thought', *PSA: Proceedings of the Biennial Meeting of the Philosophy of Science Association*, 1986 [pub. 1987], 2: 336–44; and 'Discovering Discovery: How Faraday Found the First Metallic Colloid', *Perspectives on Science*, 2006, 14: 97–121. For an overview of the use of Faraday in computer modelling, especially by Tom Addis and David Gooding, see Michael E. Gorman, *Transforming Nature: Ethics, Invention and Discovery* (Dordrecht, 1998), pp. 20–30.

For the controversy over Bošković, see Frank A. J. L. James, 'Reality or Rhetoric? Boscovichianism in Britain: The Cases of Davy, Herschel and Faraday', in *R. J. Boscovich, Vita e attivita scientifica: His Life and Scientific Work*, ed. Piers Bursill-Hall (Rome, 1993; published 1994), pp. 577–85, and 'Faraday, Michael', in *Reader's Guide to British History*, ed. David Loades, 2 vols (London, 2003), 1: 489–91. On the £20 note, see Roger Withington and B. R. James, *The New £20 Note and Michael Faraday* (Loughton, 1991).

Index

F

G

CLASSICS
A Very Short Introduction
Mary Beard and John Henderson

This Very Short Introduction to Classics links a haunting temple on a lonely mountainside to the glory of ancient Greece and the grandeur of Rome, and to Classics within modern culture – from Jefferson and Byron to Asterix and Ben-Hur.

'The authors show us that Classics is a "modern" and sexy subject. They succeed brilliantly in this regard … nobody could fail to be informed and entertained – and the accent of the book is provocative and stimulating.'

John Godwin, *Times Literary Supplement*

'Statues and slavery, temples and tragedies, museum, marbles, and mythology – this provocative guide to the Classics demystifies its varied subject-matter while seducing the reader with the obvious enthusiasm and pleasure which mark its writing.'

Edith Hall

BUDDHISM
A Very Short Introduction
Damien Keown

From its origin in India over two thousand years ago Buddhism has spread throughout Asia and is now exerting an increasing influence on western culture. In clear and straightforward language, and with the help of maps, diagrams and illustrations, this book explains how Buddhism began and how it evolved into its present-day form. The central teachings and practices are set out clearly, and keys topics such as karma and rebirth, meditation, ethics, and Buddhism in the West receive detailed coverage in separate chapters. The distinguishing features of the main schools – such as Tibetan and Zen Buddhism – are clearly explained. The book will be of interest to anyone seeking a sound basic understanding of Buddhism.

> 'Damien Keown's book is a readable and wonderfully lucid introduction to one of mankind's most beautiful, profound, and compelling systems of wisdom. His impressive powers of explanation help us to come to terms with a vital contemporary reality.'
>
> **Bryan Appleyard**

www.oup.co.uk/vsi/buddhism